# The New Nation

# A HISTORY OF US

Oxford University Press

In these books you will find explorers, farmers, cowboys, heroes, villains, inventors, presidents, poets, pirates, artists, slaves, teachers, revolutionaries, priests, musicians—the girls and boys, men and women, who all became Americans....

OXFORD

A HISTORY OF US

BOOK FOUR

# The New Nation

Joy Hakim

Oxford University Press
New York

Oxford University Press

Oxford    New York

Athens    Auckland    Bangkok    Bombay
Calcutta    Cape Town    Dar es Salaam    Delhi
Florence    Hong Kong    Istanbul    Karachi
Kuala Lumpur    Madras    Madrid    Melbourne
Mexico City    Nairobi    Paris    Singapore
Taipei    Tokyo    Toronto

and associated companies in
Berlin    Ibadan

Designer: Mervyn E. Clay
Maps copyright © 1993 by Wendy Frost and Elspeth Leacock
Produced by American Historical Publications


Published by Oxford University Press, Inc.
200 Madison Avenue, New York, New York 10016
Oxford is a registered trademark of Oxford University Press

**Library of Congress Cataloging-in-Publication Data**
Hakim, Joy.
The new nation / Joy Hakim.
p.    cm.—(A history of US: 4)
Includes bibliographical references and index.
Summary: Covers American history from Washington's inauguration until the first quarter of the 19th century, including the
Louisiana Purchase, Lewis and Clark's expedition, and the beginnings of abolitionism.
ISBN 0-19-507751-2 (lib. ed.)—ISBN 0-19-507765-2 (series, lib. ed.)
ISBN 0-19-507752-0 (paperback ed.)—ISBN 0-19-507766-0 (series, paperback ed.)
ISBN 0-19-509509-X (trade hardcover ed.)—ISBN 0-19-509484-0 (series, trade hardcover ed.)
1. United States—History—1783–1865—Juvenile literature. [1. United States—History—1783–1865.] I. Title. II. Series: Hakim, Joy.
History of US; 4.
E178.3.H22 1993 vol. 4
[E301]
973 s—dc20
[973]      93-15817
CIP
AC


5 7 9 8 6
Printed in the United States of America
on acid-free paper

*I hear America singing,* the varied carols I hear,

*Those of mechanics, each one singing his as it should be blithe and strong,*

*The carpenter singing his as he measures his plank or beam,*

*The mason singing what belongs to him in his boat, the deckhand singing on the steamboat deck,*

*The shoemaker singing as he sits on his bench, the hatter singing as he stands,*

*The wood-cutter's song, the ploughboy's on his way in the morning, or at noon intermission or at sundown,*

*The delicious singing of the mother, or of the young wife at work, or of the girl sewing or washing,*

*Each singing what belongs to him or her and to none else...*

—WALT WHITMAN,
"I HEAR AMERICA SINGING," FROM *LEAVES OF GRASS*, 1855

America, where people do not inquire concerning a stranger: What is he? but: What can he do?…The people have a saying that God Almighty is himself a mechanic, the greatest in the universe; and he is respected and admired more for the variety, ingenuity, and utility of his handiwork than for the antiquity of his family.

—BENJAMIN FRANKLIN, *INFORMATION TO THOSE WHO WOULD REMOVE TO AMERICA*

All eyes are opened, or opening, to the rights of man. The general spread of the light of science has already laid open to every view the palpable truth, that the mass of mankind has not been born with saddles on their back, nor a favored few booted and spurred ready to ride them.…These are grounds of hope for others.

—THOMAS JEFFERSON

When I hear the iron horse make the hills echo with his snort like thunder, shaking the earth with his feet, and breathing fire and smoke from his nostrils (what kind of winged horse or fiery dragon they will put into the new Mythology I don't know), it seems as if the earth had got a race now worthy to inhabit it.

—HENRY DAVID THOREAU, *WALDEN*

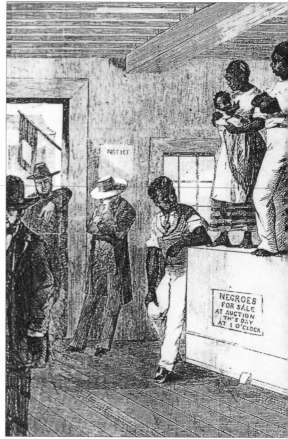

# Contents

*A page from William Clark's journal, with a sketch of the Pacific salmon.*

# PREFACE
# Getting a Nation Started

"A greater drama is now acting," wrote Washington, "...than has heretofore been brought on the American stage."

In a preface (PREFF-iss), the author is supposed to tell you what is coming in the chapters ahead. If you don't like the sound of it, you can stop reading. Well, this preface is here to tell you to read on. For this is a good book, full of stories. It is about the beginning of a nation, our nation: the United States of America.

It wasn't easy getting the country started. Mistakes were made—some big mistakes. But, mostly, we did a good job of it. Maybe it was because we had remarkable political leaders. Here are some of their names: George Washington, John Adams, Thomas Jefferson, Alexander Hamilton, James Madison, James Monroe, John Marshall. No nation has had a more impressive group of founders. They had strong ideas and strong differences. But they are only part of the story.

Imagine life in 1789. The United States has a just-written, untried constitution. A new century is soon to begin. Our young country has a president—instead of a king—and that is an idea that needs some getting used to. Never before have people written their own constitution. Never before have so many been able to vote. Never before has a nation offered its citizens complete religious liberty.

Yet, in 1789, those voting citizens are mostly white men who own property. Why should that bother some people? That is the way it is done in England. Besides, everyone knows that in the United States ordinary people can and do own land. And that is astounding in this 18th century.

It is the Constitution's words that are bothersome. The Constitution

**The first** official U.S. coin, minted in 1787, has a ring of 13 interlocking circles and the motto "We Are One." On the other side is the phrase "Mind Your Business." It is said to have been suggested by Benjamin Franklin.

9

## Democracy?

**D**o you believe in democracy—government of the people, by the people, and for the people? Today, most Americans do. But that wasn't so in times past. Democracy was an idea that took some getting used to. Many people thought it wouldn't work.

**H**enry Fearon came to America in 1818 to find out how this democratic nation was doing. Fearon had been hired by 39 English families as a kind of scout. He was to travel about and "ascertain whether any and what part of the United States would be suitable for their residence." Those families weren't moving unless they heard good things.

He began his report on the ship, before he even landed in America. He discovered a debating society on board. "Upon one occasion," said Fearon, "the question was, 'Which is the best form of government, a democracy or a monarchy?' It was strongly contested on both sides, and at length determined in favor of the former [democracy] by the [one] vote of the chairman—who was seated in presidential state on a water-cask."

When the United States was born, all but a tiny fraction of Americans lived on farms, not in cities. During the 19th century that would begin to change.

says, *We the people*. Just who are "the people"? Some Americans say that *we the people* means all people, of every color, race, and religion. Not everyone agrees. Cautious people believe the government has already gone too far with "this democracy nonsense." Others say it hasn't gone far enough. A few, who are courageous and determined, will work to bring freedom and fairness to all. The new constitution has a provision—an amendment process—that allows it to be changed.

Right away there are demands for changes. Right away 10 amendments are added. Those first 10 amendments are a bill of rights. We Americans don't want to take chances; we want to make sure that our freedoms are put down in words. The Bill of Rights gives us specific freedoms such as freedom of religion, freedom of speech, and freedom of the press.

But now, with the Bill of Rights in place, most citizens have other things on their minds. There is much to do in this newly formed nation. Ships keep bringing more and more people to America. Homes need to be built, forests cleared, and land explored.

And there are questions to be answered. No one in these United States knows what the land west of the Mississippi River is like. How wide is it? Who lives there? What plants and animals grow in that region? Someone should find out.

Why should anyone rush? Life on the land is slow-moving. Farming methods are about the same as they have been for thousands of years. Most Americans get up at sunrise and go to bed soon after the sun sets. Only the rich have watches or clocks or can afford to burn candles. Too bad we can't warn people. They are in for some big surprises. As this 18th century turns into the 19th, an industrial revolution, begun in Europe, will find its way across the sea. It will speed the pace of life.

Optimistic and productive years are ahead of us; things will go well for us Americans—except for something that is already giving our new nation a terrible, throbbing headache. This headache is caused by greed and heartlessness. Some Americans are taking advantage of other Americans. Some Americans are enslaving other Americans. Some Americans are upset about it; others don't seem to care.

Many white Americans came to this country as indentured servants. They had to work for someone else. They weren't free to leave or do what they wanted until their indenture was finished. Some were treated like slaves. So slavery doesn't seem unusual to them. Besides, slavery has been around throughout written history. The Bible talks of slavery; the Greek and Roman republics had slavery; many European, African, and Asian countries allow slavery. People are used to slavery; most people don't question things they are used to.

**Negroes for Sale.**

*A Cargo of very fine stout Men and Women, in good order and fit for immediate service, just imported from the Windward Coast of Africa, in the Ship Two Brothers.—* Conditions are one half Cash or Produce, the other half payable the first of January next, giving Bond and Security if required.

The Sale to be opened at 10 o'Clock each Day, in Mr. Bourdeaux's Yard, at No, 48, on the Bay. May 19, 1784. JOHN MITCHELL.

They are wrong not to ask questions. Slavery is terrible. American slavery is racial. The slaves are people of color—African or Native American. Slavery is economic—slaves represent money to their owners. It is very, very difficult for slaves to win their freedom.

Nevertheless, some do become free. There is a growing population of free blacks. They have jobs as carpenters, blacksmiths, farmers, cooks, and stable workers. Some are prosperous; most are not.

But as the 19th century approaches, ideas are changing. Europe's nations are beginning to outlaw slavery. One by one, the northern states outlaw slavery. According to the Northwest Ordinance, there is to be no slavery in the western territories (although some will have it anyway).

In the southern United States a way of life depends on slave labor. If

The Constitution created a new form of government, but it didn't change life for most slaves. Until 1808 it was still legal to bring slaves to the United States from Africa. And it took many more years for slaves who were already in America to win their freedom.

From its beginnings, the United States was a country where ordinary people, with no inherited advantages of birth or wealth, could own their own land and work their own farms. In the 18th-century world that was unusual.

the slave owners free their slaves, they will be giving up their wealth. People never like to do that.

Curing the headache won't be easy. There are no miracle pills around. Slavery is making some people in the North and South see things differently, and hate each other, and say so.

The American experiment in self-government may fail if this problem of injustice is not solved. How can a nation built on the idea that "all men are created equal" keep some people in chains?

It can't, of course. Our country will split apart before all its people understand that. This book is the story of America's good beginnings, and of the cruelty of slavery that will lead us to war.

# 1 The Father of Our Country

**Bringing Washington the news that he was president was one of Charles Thomson's last official acts. He didn't get a job in the new government, and he retired that year, 1789.**

George Washington was 57, and he was home, at Mount Vernon, that place he loved most. Before, when his country asked, he had left the comforts of his Virginia estate for the harshness of war. Then he left again to spend four hot months in Philadelphia, where he was needed to see a constitution written. Now, he was being asked to leave once more.

It was April 14, 1789, and Charles Thomson rode to Mount Vernon with a letter for George Washington.

Thomson—who was Irish-born—had been secretary of the Continental Congress from its beginnings in 1774. That congress was out of business. The new constitution had changed things. The confederation was finished; now there was a union of states and a new congress of the United States.

The letter that Thomson carried told George Washington that he had been elected president of that union. He had been elected unanimously—and that was important; it would not happen with any other president. It meant the government could get started without fighting over a leader. The letter said Washington was expected in New York for his inauguration. That city was to be the capital until a new one could be built.

Of course, Washington must have been proud. Martha, his wife, must have been proud of him. But he hated to leave her and Mount Vernon, especially in April. Cherry trees were in bloom; so were daffodils and tulips, and so, too, the dogwood trees, whose white blossoms

**John Adams** wrote in his diary, "This Charles Thomson is the Sam Adams of Philadelphia, the life of the cause of liberty, they say."

**Abigail Adams,** wife of John Adams—who became George Washington's vice president— wrote about the new country's leader: "He is polite with dignity, affable without familiarity, distant without haughtiness, grave without austerity, modest, wise, and good."

*Mount Vernon, wrote an English visitor, "is most beautifully situated upon a high hill on the banks of the Potomac, and commands a noble prospect of water, of cliffs, of woods, and plantations. The river is nearly two miles broad."*

Washington returned to Mount Vernon hoping to lead a quiet life. But he was besieged by neighbors asking for advice or loans, Continental army veterans who needed certificates of service, and passing strangers who just wanted to take a peek at the famous general. Washington wrote one day in his diary: "Dined with only Mrs. Washington, which I believe is the first instance of it since my retirement from public life."

*Felicity* is happiness.

floated like a layer of lace in the midst of the green woods. Washington was a farmer. In April he was thinking about spring crops and about all the chores that needed to be done on his big plantation.

But he did what his sense of duty told him to do: what he felt was best for his country. He agreed to be president. Two days later he wrote in his diary: "About ten o'clock I bade farewell to Mount Vernon, to private life, and to domestic felicity, and with a mind oppressed

with more anxious and painful sensations than I have words to express, set out for New York."

It took eight days to make the 235-mile journey. It would have been faster, but in each town citizens greeted their president-elect with a parade, or a bonfire, or fireworks, or speeches, or a ceremonial dinner, or a chorus, or sometimes all those things. So many people lined the dirt roads, and their horses' hoofs raised so much dust, that

*The story continues on page 18.*

"Long Live George Washington, President of the United States!"

On Washington's journey to his inauguration, young ladies sang and threw flowers in Trenton (left). Everywhere he was wined, dined, and fêted. At New York's Federal Hall (above), Robert Livingston, chancellor of New York State (below, third from left), swore in the new president, who promised to "preserve, protect, and defend the Constitution of the United States."

Washington said he could hardly see the countryside through the dust cloud.

Yet he was always gracious. Waving from his carriage, he saw many faces he remembered from Revolutionary War battlefields or from Valley Forge. As he approached Philadelphia he got out of his carriage, mounted a white horse, and rode toward the city. The parade of horsemen that followed him grew longer and longer.

At a bridge that spanned the Schuylkill River at Gray's Ferry, Washington couldn't help but be dazzled. A grove of laurel and cedar trees seemed to be growing out of the water. At each end of the grove, tall, leafy arches were covered with flags, ribbons, and flowers. It was the work of his friend and fellow soldier, the inventive painter Charles Willson Peale. Peale's 15-year-old daughter, Angelica, was hiding in the artistic shrubbery. As Washington passed under an arch Angelica pulled a lever and a laurel wreath fell and rested above the hero's head. Modestly he rode on, after kissing the girl.

At Trenton he crossed another bridge decorated in his honor, this one with patriotic banners. He remembered being in Trenton during the Revolutionary War with his half-frozen soldiers. Now a chorus of women and girls sang and threw flowers in his path. "Strew your hero's way with flowers," they sang. He called it an "affecting moment." That night, after yet another public dinner, he took time to write a thank-you letter to the women and girls.

When he finally arrived in New York—rowed across the Hudson River from New Jersey on a barge decorated with streamers—church bells rang, cannons roared, and people cheered until they were hoarse.

Ever modest, Washington thanked the crowd and said, "After this is over, I hope you will give yourselves no further trouble, as the affection of my fellow citizens is all the guard I want."

Six days later they were still cheering. It was the day of his inauguration—April 30, 1789. Washington wore a plain brown suit made of American cloth, stood on the balcony of Federal Hall overlooking Wall Street, bowed to the great crowd below, put his hand on his heart, and took the oath as president.

Afterwards, he and the members of Congress walked up Broadway to spired, stately St. Paul's Episcopal Church. There, under a blue-sky ceiling, the new president prayed for guidance, for the young republic, and for himself.

## Kings and Revolution:

While George Washington was president, other nations were ruled by kings, emperors, and tsars (ZARS). Remember, the birth of the United States was a world-shaking event. We were the first modern nation to form a people's government, to write our own constitution, and to elect our own leaders.

Most Americans in the 18th and 19th centuries knew a lot about kings, and didn't want one in the United States. Have you ever wondered why? Read on and see if you'd like a king for a ruler.

Charles IV was king of Spain when George Washington was president of the United States. Charles got up each morning at five o'clock and said prayers. Then he went off to visit the royal stables and the royal armory (which was where the royal guns were kept). Sometimes he went to the royal art gallery and admired the royal paintings. He always made time for the queen and his children to come and kiss his royal hand. Exactly at noon he sat down to a large, royal dinner. He ate alone—no one was exalted enough to eat with the king.

Every afternoon, rain or shine, the king went hunting. Now a king doesn't just go off with a rifle over his shoulder. It isn't that simple. Six coaches full of companions went with Charles each day. Three hundred men went into the woods to drive the game toward the hunters. In all, it took 700 men and 500 horses to see that the king hunted properly.

This, of course, was expensive—but what are a king's subjects for?

*King Charles IV of Spain and family.*

Spanish taxes were high, but the Spaniards could boast that their king was one of the best shots in Europe. In the evening, the king tended to the nation's business. That was just before he played his violin. Then King Charles went to bed in order to be up early to say his morning prayers.

King Louis XVI and Queen Marie Antoinette were the rulers of France in 1785, when Thomas Jefferson arrived as America's ambassador. France's King Louis XVI didn't eat alone. Three hundred and eighty-three men, including nobles, waited on him and presented each morsel of his food in a splendid, ceremonial fashion. It took four people just to serve the king a glass of water.

Democratic-minded Jefferson was horrified by all this extravagance. He wrote a letter to George Washington. "I was much an enemy to monarchy before I came to Europe," said Jefferson. "I am ten thousand times more so since I have seen what they are."

It cost the French people a great deal of money to pay for all the pleasures of their king and queen. Taxes were very high. Supporting the royal family was making the people poor.

Finally, the French got fed up. (Actually, they weren't fed up—many of them were hungry.) The French people knew all about the American Revolution; they decided to have one of their own. The French Revolution began on July 14, 1789, when angry citizens stormed a horrid old prison called the Bastille (bass-TEE). The king and the nobles had been putting political prisoners in the Bastille and forgetting them there. After the revolution, the key to the Bastille was sent to President Washington.

The French people wanted to form a democratic government. After the revolution, the king and queen said that they would reform the government, but, secretly, they were planning to hold on to power. That was a mistake. They lost their heads. (Yes, their heads.)

Then the revolution went awry (uh-RYE—which means it got off track, sour, wrong). Some brutal leaders began killing people and couldn't seem to stop. A strong man—a powerful French general named Napoleon Bonaparte—took over. Soon he was fighting most of Europe—until he was defeated and imprisoned on the island of Elba (which is in the Mediterranean Sea, near Italy). Finally, France did get a democratic government.

Naturally, people in America found all this very interesting. So did kings and queens in the rest of the world. They were shivering in their shoes. America's Tom Paine didn't make them feel any better. Paine wrote a book called *The Rights of Man*. It urged the English to overthrow their king too.

*Louis XVI of France before he lost his head.*

When the French Revolution turned nasty, many French people wanted to leave their country. Some had been supporters of the king. Some were revolutionaries. Some were afraid they might have to fight in Napoleon's army. Many of those people came to America instead.

*The storming of the Bastille on July 14, 1789, to release its prisoners, was the opening act of the French Revolution. Today France celebrates that date as we Americans celebrate July 4.*

19

# 2 About Being President

**Washington himself sketched the design he wanted for the doors of his presidential carriage.**

*Precedent* (PRESS-ih-dent) and *president* (PREZ-ih-dent). They sound similar, but their meanings aren't the same. Say them aloud and hear the difference.

> **George Washington**
> PRESIDENT, 1789–1797

No one could tell George Washington how to be president. No one had done the job before. Washington knew that whatever he did would set a precedent. That means he would be the example and other presidents would follow his lead.

The Constitution outlined the basic tasks of the president, but it didn't go into details. George Washington had to decide many things himself.

As always, he did his very best. He didn't want the president to be like the English king, but he did think it important that the president be grand. He wanted people to look up to the president and respect and admire him.

So Washington acted with great dignity and rode about in a fine canary-yellow carriage pulled by six white horses whose coats were shined with marble dust, whose hoofs were painted black, and whose teeth were cleaned before every outing.

When President Washington held official receptions he wore velvet knee breeches, yellow gloves, silver buckles on his shoes, and a sword strapped to his waist. He used his coach to tour the country. He wanted

**Washington liked things just so. He also chose his carriage's paint and seat fabric.**

Americans to meet their president.

As president he was head of the executive branch of our three-branch government. (The other two branches are the legislative, which is Congress, and the judicial, which is the courts.) Washington knew he couldn't make all the decisions of the executive branch by himself. So he appointed advisers. Most of those helpers were called secretaries: secretary of state, secretary of the treasury, and so on. All together they were known as the cabinet.

The president and his first cabinet. Left to right, Washington, Knox, Hamilton, Jefferson, and Randolph. Not in the cabinet, but just as important, was first speech writer Madison (below).

Washington picked the very best people he could find. To help with foreign affairs, he picked an American who had been Virginia's governor and had lived in France and knew a lot about foreign nations. Can you guess who he was? Well, George Washington named Thomas Jefferson as his secretary of state.

You can't run a country without money. Since the days of the Revolution, when the states first united, they had had money problems. Washington needed a good man as a financial adviser. He named Alexander Hamilton as secretary of the treasury. Hamilton organized the nation's monetary system. Some people think that Alexander Hamilton was the best secretary of the treasury ever.

To head the army and navy, Washington chose his old friend Henry Knox. Remember Knox the Ox? He was the general in charge of artillery during the Revolutionary War. Washington named him secretary of war.

John Adams, who had been elected George Washington's vice president, was also a cabinet member. Washington completed the cabinet when he appointed Virginia's governor, Edmund Randolph, as attorney general.

When he needed help writing a speech, President Washington turned

to a congressman who had one of the finest minds in American history: James Madison. (And when Congress wished to address the president, guess who wrote the message? James Madison. So Madison was writing and answering the same messages!)

Altogether, Washington had about 350 people help him manage the new government. That was only about a hundred more people than he supervised at his plantation home, Mount Vernon.

Cockfighting, an ancient betting sport, began in Asia and reached Europe in the fifth century B.C.E. It was no sport for the roosters—they were usually fitted with metal spurs and fought to the death. Here, two men seem to be fighting, too.

Almost as soon as the new government got started, something happened that Washington hadn't expected. His two top advisers argued with each other. They really argued. Jefferson and Hamilton had ideas that clashed. They found it hard to compromise. In those days people sometimes watched cockfights, and so when Jefferson wanted to describe himself and Hamilton, he said, "Hamilton and myself were daily pitted in the cabinet like two fighting cocks."

Both were brilliant men. Both were patriots who wanted to do their best for their country. They just disagreed on what was best. Did they ever disagree! In fact—this is interesting—political parties developed because of that disagreement.

Here, to **pit** means to set against each other. It doesn't have anything to do with the pit inside a cherry or a peach. Cockfights were staged in pits.

The country didn't begin with parties like today's Democrats and Republicans. The Founding Fathers—the men who wrote the Constitution—didn't realize that parties would develop. Washington didn't like the idea at all. He called them "factions" and warned against them. "The spirit of party," said the president, "agitates the community with ill-founded jealousies and false alarms."

But people just don't think alike. That's what makes politics and life interesting. James Madison understood that. Madison knew that it was dictators who usually try to force all people to think alike. Dictatorships are one-party governments.

Madison believed that in a democracy factions should be encouraged. He thought the more the better. He said they would balance each other and then no one group could become too strong and take control of the government.

## Early Bird

It was 1792, and Jean-Pierre François Blanchard was up in the air with a note in his pocket written by President George Washington. The note explained who he was, so that when he landed he would not be mistaken for an alien from outer space. Blanchard was doing something several Americans had tried, but none successfully. He was flying in a balloon.

*Blanchard hoists Stars and Stripes over New Jersey.*

from Dover, England, to Calais, France. For scientific-minded Americans in the Age of Enlightenment, this day was full of wonder and excitement. Was it possible that a man could actually rise into the sky and keep company with the birds?

Blanchard and his balloon drifted without any problems for 46 minutes and then descended 15 miles away near Woodbury, New Jersey. Fascinated observers helped pack up the deflated balloon, load it onto a wagon, and haul it back to Philadelphia. Then the triumphant skyman called on the president.

Blanchard gave Washington the flag he'd taken aloft: on one side was the American flag, on the other the French tricolor.

Thousands of people, including George Washington and Thomas Jefferson, watched as Blanchard lifted off from the courtyard of the Walnut Street Prison in Philadelphia. Blanchard had already made 44 successful balloon flights in Europe. In 1785 he had even crossed the English Channel

**Nowadays,** the cabinet (look on page 21 for the details of Washington's cabinet) also includes secretaries of these departments: interior; labor; agriculture; commerce; health and human services; housing and urban development; transportation; energy; veterans' affairs; and education.

**Do words** and titles matter? Today the secretary of war (Henry Knox's old job) is called the secretary of defense. If you were voting in Congress, would you be more likely to give money to a war department or a defense department? Could the title *secretary of defense* be a euphemism (YOO-fuh-miz-um)? That means a pleasant or inoffensive word or name substituted for an unpleasant or disturbing one.

23

## The Sense of America

*Hamilton published an attack on Jefferson saying that he was disloyal to the Constitution. Jefferson wrote a letter to President Washington answering that charge.*

**N**o man in the United States, I suppose, approved of every title in the constitution; no one, I believe, approved more of it than I did...my objection to the constitution was that it wanted a bill of rights, securing freedom of religion, freedom of the press, freedom from standing armies, trial by jury, and a constant habeas corpus act. Colonel Hamilton's was that it wanted a King and House of Lords. The sense of America has approved my objection, and added a bill of rights, not the King and Lords.

# 3 The Parties Begin

"Men by their constitutions," said Thomas Jefferson, "are naturally divided into two parties."

Those two fighting cocks—Jefferson and Hamilton—had ideas that needed balancing. They helped found the country's first political parties. They respected, but didn't understand, each other.

"Mr. Jefferson," wrote Hamilton, "is at the head of a faction decidedly hostile to me and...dangerous to the union, peace, and the happiness of the country."

Jefferson replied that Hamilton's ideas "flowed from principles adverse to liberty, and...calculated to undermine and demolish the Republic."

Whew! Those are strong words. "Dangerous," "hostile," "adverse to liberty"—did they really mean it? These were men who had built the country together. What was going on?

To put it simply: they disagreed about power and who ought to have it. It was that old conflict that had kept everyone arguing when the Constitution was being written.

Jefferson and Hamilton were both concerned about liberty and about power. How do you balance the two? How do you guarantee freedom? How do you create a government that can keep order and make sure that government doesn't oppress people? How strong should the government be?

Hamilton believed the government should be strong. If the government was to work for all the people, instead of just those with the loudest voices, it needed to be powerful. Hamilton thought that government should be run by aristocratic leaders, that is, by the prosperous, well-educated citizens who he thought had the time and talents to best run a country. He feared the masses. He said they sometimes acted like sheep, thoughtlessly following a leader.

**Here is** an old saying the Founders had probably heard: *Liberty for the whale means death for the minnow.* What does that mean? How does it relate to a free nation?

But Hamilton was also wary of the rich. He thought they often acted out of self-interest—that means they did what was good for them. Hamilton knew the government needed checks and balances so no group could gain control.

"Give all power to the many," wrote Hamilton, "and they will oppress the few. Give all power to the few, they will oppress the many. Both therefore ought to have power, that each may defend itself against the other."

Thomas Jefferson feared powerful government. It was justice and liberty for the individual that concerned him. He saw a strong, centralized government as a possible enemy of individual liberty. Jefferson had been in Europe and had seen kings in action: he hated monarchies. He feared a king-like president.

Jefferson had faith in ordinary people. He thought they could govern themselves—if they were educated. And so he wrote a plan for public schools and colleges. He wanted an amendment to the Constitution that would provide for free education.

Because of the differences in ideas, it became clear that political parties were needed. Hamilton's followers formed the "Federalist Party." Jefferson's followers were called "Democratic-Republicans," or sometimes just Republicans.

Now this is confusing, so pay attention. The Federalists and Republicans were not like our Democrats and Republicans—but they were the beginnings of today's party system. This is what is confusing: Jefferson's Republican Party was not like today's Republican Party. Actually, it was the parent of today's Democratic Party. (The modern Republicans got started later with a president named Abraham Lincoln.)

Jefferson and Hamilton were both good men, and the ideas of each of them have been important in our country. On most issues (but not all), Hamilton was a "conservative"

**Jefferson fought for freedom of the press. That meant papers would be free to criticize him. Here, Washington is the hero with the halo over his head; Jefferson is the villain with a smoky black candle as his symbol. How do you think T.J. felt about that?**

What are **masses**? "Mass" is the scientific term for any quantity of matter, tiny or huge. That wasn't the definition Hamilton had in mind. Another meaning of "mass" is a great many people. *The masses* came to mean the workers and ordinary folk in a country. Hamilton didn't think himself part of the masses. Although he was a poor boy, he became rich and well educated.

**Jefferson didn't** get his free education amendment. His plan for education in Virginia was turned down in the Virginia General Assembly.

LOOK ON THIS PICTURE,    AND ON THIS.

THE PROVIDENTIAL DETECTION

**Jefferson is attacked for sympathizing with the French revolution. The Federal eagle is rescuing the Constitution before Jefferson burns it. Was this the true state of affairs?**

One dictionary's definition of **conservative** is "tending to oppose change." Conservatives like to hold on to ideas from the past. **Liberal** is defined as "having political views that favor civil liberties, democratic reforms, and the use of governmental power to promote social progress." But these words have had some very different meanings at different times.

and Jefferson a "liberal." Have you ever heard people argue about conservatism and liberalism? Well, if you haven't, you will. That argument almost tore the country apart in 1800, and it continues today. Which is better: conservatism or liberalism? I think it is the tension and the compromises between those two ideas that have helped make this country great. We need Hamiltonians, we need Jeffersonians, and we need to have them work together.

Which is just what has always happened in America. That is not true in many other nations.

In some countries, people who speak out against the government are put in jail, or even killed. Members of the losing party are thrown out of the country, or even killed. That doesn't happen in America. Here, since the time of President George Washington, winners and losers have always agreed to work together—as Thomas Jefferson and Alexander Hamilton did. What does that mean for you? Do you have to be afraid of being on the side of the losing party? Can you speak out for an unpopular cause? Of course you can, you're an American.

## Opposite Sides of a Penny

Jefferson said, "The many!"
Hamilton said, "The few!"
Like opposite sides of a penny
Were those exalted two.
If Jefferson said, "It's black, sir!"
Hamilton cried, "It's white!"
But, 'twixt the two, our Constitution started working right.

Hamilton liked the courtly,
Jefferson liked the plain,
They'd bow for a while, but shortly
The fight would break out again.
H. was the stripling Colonel
That Washington loved and knew,
A man of mark with a burning spark
Before he was twenty-two.

He came from the warm Antilles
Where the love and the hate last long,
And he thought most people sillies
Who should be ruled by the strong.
Brilliant, comely and certain,
He generally got his way,
Till the sillies said, "We'd rather be dead,"
And then it was up to J.

He could handle the Nation's dollars
With a magic that's known to few,
He could talk with the wits and scholars
And scratch like a wildcat, too.
And he yoked the States together
With a yoke that is strong and stout.
(It was common dust that he did not trust
And that's where J. wins out.)

—STEPHEN VINCENT BENÉT, "ALEXANDER HAMILTON"

26

"No other American statesman has personified national power and the rule of the favored few so well as Hamilton, and no other has glorified self-government and the freedom of the individual to such a degree as Jefferson." It is a conflict that runs through American history and continues today.

*Alexander Hamilton*

# Money, Money, Money, Money

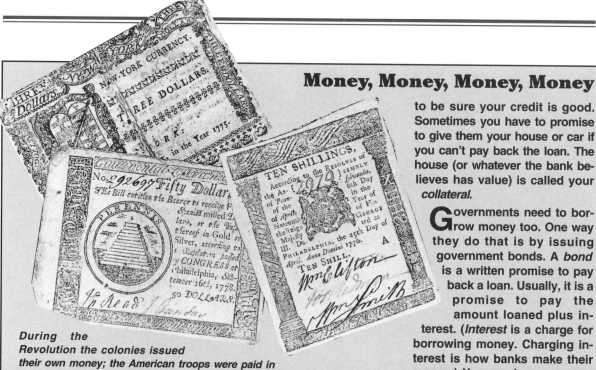

During the Revolution the colonies issued their own money; the American troops were paid in "Continental dollars." But the government had little gold or silver to back the bills; soon any valueless object was said to be "not worth a Continental."

Lots of people argue about money, so you may not be surprised to learn that Hamilton and Jefferson did that too. They had different visions of the way they wanted the United States to grow. Hamilton wanted to encourage business and industry. Jefferson hoped to keep America a nation of farmers and landholders.

But the world was changing, and, whether Jefferson liked it or not, cities and factories were on their way. So was a *money economy.*

In early America, most people were self-sufficient farmers and had little use for money. They bartered—traded—for what they needed. About the time the United States became a nation, we were turning into a *capitalistic* society.

Here are some words to help you understand about capitalism: *capital* is money, or any goods or services that can be turned into money. If your family owns a house or car that can be sold for cash, that is part of your family's capital, along with any money you have in the bank and in your pocket.

The grease that lubricates the wheels of a capitalistic economy is *credit.* Credit is borrowing power. If you want to start a business you will probably need to borrow money to do so. (Most businesses are started that way.) The bank, or whoever lends you the money, needs to be convinced you will pay back the loan. They need to be sure your credit is good. Sometimes you have to promise to give them your house or car if you can't pay back the loan. The house (or whatever the bank believes has value) is called your *collateral.*

Governments need to borrow money too. One way they do that is by issuing government bonds. A *bond* is a written promise to pay back a loan. Usually, it is a promise to pay the amount loaned plus interest. (*Interest* is a charge for borrowing money. Charging interest is how banks make their money.) You can buy a government bond. You will be lending money to your government and you will earn interest too.

Just like people, some governments pay their debts promptly and have good credit. Some governments *default* on their bonds. That means they don't pay what they owe. Naturally, they have difficulties the next time they want to borrow money.

In the time of George Washington, when the United States was just beginning, it owed a lot of money. The Congress of the Articles of Confederation had borrowed from citizens and from other nations in order to fight the Revolutionary War. The debt was huge and it was not being paid off properly. The bondholders were not getting interest payments. They could not get their money back for their bonds.

Some Americans thought that was all right. They thought it was unfair for the new nation to be stuck with the debts of the old government. Some people suggested that the new United States government ignore those old debts.

**M**any former soldiers and farmers and everyday citizens had lent the government money. That made them *investors*. They held government bonds. Those citizen-investors believed their bonds were worthless. There were rumors that the government would not pay the debt. Many of those investors sold their bonds for much less than their *face* (promised) value. Some people who held $100 government bonds sold them for $25. They thought they were lucky to get anything for them.

The people who bought the bonds were *speculators*. They were taking a chance. Many of the speculators were wealthy, so they could afford to take a chance.

That was the situation when Alexander Hamilton became secretary of the treasury. The government debt was $64.12 million. It would be very difficult to pay that enormous amount of money. "Couldn't the United States start with a clean slate?" many people asked.

Jefferson and Madison believed the new government should not be responsible for the mistakes and debts of the old. They didn't want to see speculators get rich.

Hamilton disagreed. He said, "States, like individuals, who observe their engagements, are respected and trusted, while the reverse is the fate of those who pursue an opposite conduct." Hamilton decided to pay off the debt. He got Congress to do it. Jefferson and Madison were furious.

**B**ut when Hamilton left office (when he stopped being treasury secretary), in 1795, the United States had a fine credit rating. Everyone wanted to buy U.S. government bonds, because they knew they could trust the new nation. Our country was standing on sound financial feet.

**S**ome people call capitalism a *free market* economy. That sounds like a place where people can do business without government regulations. But every modern country has rules and regulations for business. You wouldn't want a no-rules-at-all country. A dishonest grocer could cheat and make his scales say two pounds when he was selling only one pound. If ingredients weren't listed on food and cosmetic packages, anything might be inside, and you'd never know—until you got sick. But in a free-market economy you can choose your own business; you can usually buy and sell where you wish; you have a great deal of freedom.

*The first American bank, Philadelphia's Bank of the United States, cost so much to construct that the builders had to finish the sides in brick instead of marble.*

# 4 A Capital City

**Pierre L'Enfant, planner of the Federal City, was a Revolutionary War veteran who had wintered at Valley Forge.**

**Benjamin Banneker was the first black man to receive a presidential appointment—but he still could not vote.**

*You know another famous American who published an almanac. Who was he?*

Hamilton and Jefferson did agree on one thing. They agreed that a new capital city was needed. New York just wasn't right, so in 1790 the capital was moved to Philadelphia. But Philadelphia wasn't right either.

Each state wanted the capital city for itself. Which state would get it?

There was a sensible answer: a new city would be built and it would not be in any state—then there would be no state jealousy. It would have its own piece of land in a special place to be known as the "District of Columbia." (I don't have to tell you who that was named after, do I?)

The capital was called the Federal City. George Washington picked the site for it on the Potomac River. It seemed a good choice—right in the middle of the country. (Middle of the country? Yes, in 1790.) Congress approved.

Now the city needed planning. Thomas Jefferson had visited and studied cities and gardens in Europe. He was fascinated by architecture and landscape design. He had many ideas for the new city. But professionals were needed to set to work. To begin, the land needed surveying. Benjamin Banneker was hired to do much of that.

Banneker was a free black and a farmer. Some people called him a genius. When he was a boy he went to a Quaker school in Maryland, but after he left school he kept studying on his own. He studied mathematics and astronomy. He played the violin. He studied and wrote about bees and locusts. He designed a clock and built it all of wood. From 1792 until 1802 he published a popular almanac (a book about events of the year). Those are just a few of the things he did.

Because he was black, Banneker could never enjoy the full rights of citizenship. He knew that was wrong and he tried to do something about it. Thomas Jefferson had said all men are created equal, so Banneker wrote to Jefferson and asked him to help end "that train of absurd and false ideas and opinions which so generally prevail with respect to us [blacks]."

Jefferson answered Banneker and said of his letter: "I consider it a document to which your whole color had a right for their justification against the doubts which have been entertained of them." (And that means, in plain English: anyone who has doubts about black people should read this brilliant letter.)

Jefferson agreed with what Banneker said, but he didn't do much to change the "absurd and false ideas" held by many in the nation. Perhaps he thought he couldn't.

Banneker worked hard to make his country better for all people. He proposed that the cabinet

Hoban's design (inset top) won the President's House contest, but Jefferson's (with a dome) may have influenced other Washington buildings, including the Capitol (left, before its dome was built). Below, L'Enfant's grand plan.

have a secretary of peace as well as a secretary of war. He worked for free public education and to end capital punishment.

The man who actually planned the city was another genius. Pierre Charles L'Enfant (lahn-FAHN) studied art and engineering in Paris and then came to America and fought in the Revolutionary War. After that, he helped plan a New York dinner party for 6,000 people! It was a big party to celebrate New York's ratifying of the Constitution, and it was a great success.

L'Enfant's first big architectural job came when he redesigned New York's old City Hall and turned it into a headquarters for the new government. It was named Federal Hall. But it was as the designer of the new nation's capital that he became famous. L'Enfant put a broad, grassy mall in the center of the city-to-be. Then he laid out rectangular city blocks and cut through them with broad, spoke-like avenues. He decided that the House of Congress, called the Capitol, was to be on a hill (Capitol Hill) overlooking the mall and the Potomac River. From the President's House there would be another view of mall and river. It was a splendid design.

A contest was announced to find the best plan for the President's House. One person who entered the contest put a false name on his entry. He was an amateur architect (a man who loved to design buildings), but he didn't want the judges to know his real name. Can you guess who it was? The secret competitor was Thomas Jefferson. He didn't win the competition. James Hoban, an Irishman, won and became architect of the White House.

Another contest was announced. This was for the Capitol. Some strange entries arrived: a bridge builder whose bridges collapsed sent a design. So did a carpenter who seemed to like windowless rooms. The winning entry came from William Thornton, who was born in the British Virgin Islands and had studied to be a doctor in Edinburgh, Scotland. George Washington said Thornton's design combined "grandeur, simplicity, and convenience."

It took forever to build the Capitol. Other architects were soon adding their ideas to the winning entry: Frenchman Etienne Hallet was one; Thomas Jefferson was another; English-trained Benjamin Henry Latrobe was yet another. The Capitol was still under construction when our 16th president took office. (His name was Abraham Lincoln.)

From its beginnings, the United States government looked for people with good ideas and didn't worry about their differences. It was appropriate that a Frenchman, an African American, a Scotsman, an Englishman, and an Irishman all helped to make our nation's capital beautiful.

Notice those words: *capitol* for the building and *capital* for the city. How will you remember which is which? Well, one way to do it is to think of the round dome on top of the Capitol and think of the letter *o*, which is round.

All those *cap-* words come from the Latin word *caput*, which means head. Does *capital punishment* have anything to do with heads?

32

# 5 Counting Noses

In this chapter you need to pretend that you are married. Now pretend you have a big family—five boys and five girls. Keep pretending, because those children of yours are growing up. Each one gets married and has a family of his or her own. Then your grandchildren begin to have families too.

By this time there are so many members of your family that you've lost track of them all. You don't even know how many grandchildren and great-grandchildren you have, or where they all live.

**The first census questioned only men aged 21 years or older about their families. In 1790 it was rude to ask about a lady's age or circumstances.**

How can you make family plans? Well, of course, you can't.

So you decide to find and count your family. And that was exactly what the new country did in 1790. After all, how can you run a nation when you don't know how many citizens you have?

In 1790 the United States took its first census. It counted all its citizens. We've been doing that every 10 years since 1790.

Almost 4 million of us were counted in 1790—3,929,214 to be exact. Six hundred and ninety-seven thousand, six hundred and eighty-one of us were slaves. No one counted the Indians. We do know that half of the land claimed by the United States government was held by Indians.

33

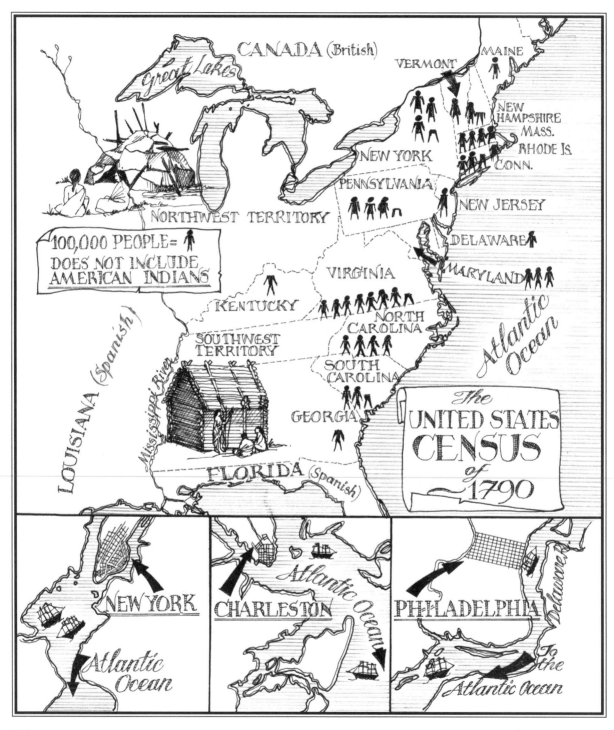

CANADA (British)

Great Lakes

VERMONT

MAINE

NEW HAMPSHIRE

MASS.

RHODE IS

CONN.

NEW YORK

PENNSYLVANIA

NEW JERSEY

DELAWARE

NORTHWEST TERRITORY

100,000 PEOPLE =
DOES NOT INCLUDE
AMERICAN INDIANS

MARYLAND

VIRGINIA

Atlantic
Ocean

LOUISIANA (Spanish)

KENTUCKY

NORTH
CAROLINA

SOUTHWEST
TERRITORY

SOUTH
CAROLINA

Mississippi River

GEORGIA

The
UNITED STATES
CENSUS
of
1790

FLORIDA (Spanish)

NEW YORK

Atlantic
Ocean

CHARLESTON

Atlantic Ocean

PHILADELPHIA

Delaware

To
the
Atlantic Ocean

There was someone special who didn't get counted in 1790. He was Benjamin Franklin. Ben died that year. He'd been around so long—almost all of the 18th century—that it was hard to think of America without him. Franklin had helped the country grow, become independent, and write its own constitution. He died in peace.

Ben was a city person—he lived in Boston, Philadelphia, and London—and that was unusual. In 1790, most Americans—95 percent of us—lived on farms. Just 3 percent of America's citizens lived in the six largest cities: Philadelphia, New York, Boston, Charleston, Baltimore, and Salem. And those who lived in the city often had chickens, a cow, and a vegetable garden.

The city dwellers were very important even though they were few in number. People exchange ideas in cities. Markets are in cities. Newspapers and books are published in cities. Farm people travel to cities to trade goods and thoughts. So city people influence others.

Look at the map, and you will notice that all the big cities in early America were ports. Shiploads of new people—immigrants—were sailing into those ports almost every day. Most of those immigrants moved on. They cleared farmland or helped build more new cities. And they had children, lots of children. America's birth rate was astonishingly high, twice that of England, and the death rate was low. This land, with its fresh air, pure water, fish, game, and fertile land, was a healthy place to live. In America in the 19th century you could have 10 children (many couples did), and there was food and opportunity for all of them.

So it was no surprise when each new census showed huge population gains. For almost a century, the United States doubled its population about every 24 years. Think about the arithmetic of that. When you keep doubling you get into big numbers quickly.

The census of 1800 counted 5.3 million Americans. One million of us were black, and nine of every 10 blacks were slaves. Philadelphia, in 1800, had 70,000 citizens. (Remember, it had 40,000 in 1787 when the Constitution was written.) In New York there were 60,000 people; in Boston, 25,000.

The new country was much larger than just the 13 states. When the British lost the Revolutionary War and signed a peace treaty, England gave up land that stretched all the way to the Mississippi River. (If you crossed the Mississippi, as Daniel Boone did, you were in a foreign land. You were in Spanish territory until 1800. Then the land was taken by France. And then—keep reading and you'll find out who got it next.)

By 1800, almost one million people lived in the frontier region

**In 1790** there were fewer than 4 million Americans. By 1890 there were 63 million Americans. What is the population today?

**Creek and** Yuchi Indians lived in Tennessee when the Spanish explorer de Soto marched through in 1540. French explorers followed, then English, then Daniel Boone, and then Elisha Walden, who came with long hunters (they got the name because they spent long periods of time in the wilderness). Some independent mountain folk established a free state in the northeastern section and called it Franklin (after Benjamin Franklin), but most settlers wanted to be part of the United States. In 1796, Tennessee became the 16th state.

**In 1800,** when Americans said they were heading west, they usually meant they were on their way to Ohio. In 1800 Ohio had 45,365 non-Indian inhabitants; 40 years later it had 1,519,467. Ohio became the 17th state in 1803. In 1840, Ohio had the third largest population of any state, behind New York and Pennsylvania.

**City life brought Americans into contact with people different from themselves, as this Philadelphia street scene of an oyster vendor shows.**

between the Appalachians and the Mississippi. The bluegrass state of Kentucky (number 15 in the new nation) had 220,955 settlers, which was more people than lived in the whole of New Jersey (one of the first 13). Ohio, which became a state in 1803, had 45,000 inhabitants.

On the frontier—beyond the mountains—people often lived in crude huts and wore leather clothes adapted from Indian designs. Their lives were hard, free, and sometimes surprising.

A 19th-century visitor from France described a frontiersman as "a highly civilized being, who consents for a time to inhabit the backwoods, and who penetrates the wilds of the New World with the Bible, an axe, and some newspapers." On the frontier it didn't matter who you were; it was what you could do that counted.

Boys and girls on the East Coast—those with white skins—lived much as their European cousins did. But there was an important difference: people here didn't have long traditions to guide them (as most people on other continents did). So Americans learned to think for themselves.

That was true for black boys and girls, too. Their world was radically different from that of their African cousins. They learned new ways, and, at the same time, hung on to some of their old traditions. It wasn't easy to do.

**Before good roads were built, rivers were the roadways to the frontier lands; cities such as Pittsburgh (below) grew up as settlers moved west.**

# 6 The Adams Family Moves to Washington

**"If a woman does not hold the Reins of Government,"** said Abigail Adams, **"I see no reason for her not judging how they are conducted."**

Our second president was the first president to live in the Federal City. He was John Adams, that solid thinker from Massachusetts who had helped convince Thomas Jefferson to write the Declaration of Independence.

The Federal City wasn't much of a place when John Adams became president. The city was still being built. The new president and his wife, Abigail, headed south from Philadelphia looking for the capital. They got lost in the woods.

Now you may remember that Abigail Adams was a strong woman who liked to write letters and who said whatever was on her mind. In a letter she wrote to her daughter she described the new city and the President's House (it would later be called the White House):

> Woods are all you see, from Baltimore until you reach the city, which is only so in name...there are buildings enough, if they were compact and finished, to accommodate Congress and those attached to it; but as they are, and scattered as they are, I see no great comfort for them. The river, which runs up to Alexandria, is in full view of my window, and I see the vessels as they pass and repass. The house is upon a grand and superb scale, requiring about thirty servants to attend...and perform the ordinary business of house and stables....To assist us in this great castle, and render less attendance necessary, bells are wholly wanting, not one single one being hung through the whole house, and promises are all you can obtain.

Despite all those woods, the president and his wife didn't have

**After picking** his way through the muddy, unpaved streets of the new capital, Gouverneur Morris remarked that the Federal City might be all right for the future, but that he was not posterity. (What is *posterity*?)

A ***drawing room*** has nothing to do with drawing pictures. It is short for *withdrawing room,* because it was the room the ladies withdrew to after dinner while the men drank and smoked in the dining room.

enough wood for their fireplaces, and there was no one to cut it. John Adams wouldn't have slaves; there weren't many laborers in the new city; and wood was very expensive. Abigail complained that they couldn't afford wood on the president's salary and that the President's House was damp. (There were no furnaces then; fires in the fireplaces kept houses warm and dry.)

The main stairs were not yet in, said the First Lady.

> We have not the least fence, yard, or other convenience without, and the great unfinished audience-room I made a drying-room of, to hang up the clothes in.

If the city had been built in New England, she wrote to her daughter, naturally, *many of the present inconveniences would have been removed.* Still, Abigail found many things to admire.

> Upstairs there is the oval room, which is designed for the drawing-room, and has the crimson furniture in it. It is a very handsome room now; but, when completed, it will be beautiful....[The city] is a beautiful spot, capable of every improvement, and, the more I view it, the more I am delighted with it.

The Federal City was not far from George Washington's home at Mount Vernon. The first president watched as the new city and the second president got settled. Then, in 1799, something terrible happened. George Washington got sick and died. It happened suddenly,

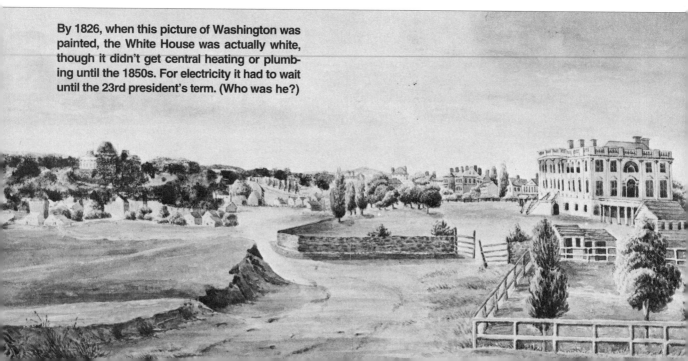

By 1826, when this picture of Washington was painted, the White House was actually white, though it didn't get central heating or plumbing until the 1850s. For electricity it had to wait until the 23rd president's term. (Who was he?)

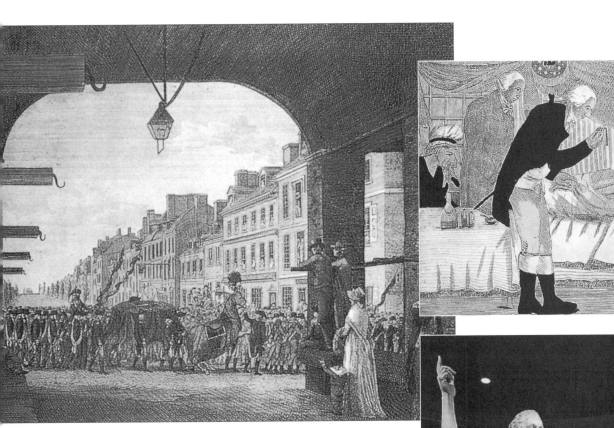

"I die hard, but I am not afraid to go," said George Washington on his deathbed in 1799 (above right). The whole country mourned (above, the memorial procession in Philadelphia). Many Americans saw Washington almost as a god or saint, an attitude that embarrassed him greatly. What would he have thought of the statue (right) that portrays him as Zeus?

just two weeks before the start of the new century that George Washington had hoped to see. All the country wept.

Abigail Adams wrote to her daughter that she had received an invitation to visit Martha Washington at Mount Vernon. Of course she would go and pay her respects to the widow.

Richard Henry Lee, Washington's good friend and fellow soldier, said what citizens everywhere were thinking. He said Washington was "first in war, first in peace, first in the hearts of his countrymen."

Everyone agreed that the nation's new home—the muddy and unfinished Federal City—should be named Washington to honor its first president.

# 7 About President Adams

We have presidential buttons and bumper stickers. In Adams's day he could find himself on jars and jugs.

**"The times,** Madam, have made a Strange Being of me," said John Adams to a friend, historian Mercy Otis Warren. He was, he admitted, "an irritable Fiery Mortal...as proud as a Caesar. But an honest man in all and to the Death."

Joseph Ellis (a 20th-century writer) says that Adams "has come to be regarded by historians as the most engagingly human member of America's founding generation." *Engaging* means charming or appealing. What was it that made Adams human, too?

**John Adams**
PRESIDENT, 1797–1801

John Adams was a great man, but he was just passable as president. Now that is my opinion; you are free to disagree. Some historians do. Adams was a fine person—honorable and thoughtful. So was his wife, Abigail, and so, too, was their brilliant son John Quincy. Benjamin Franklin, who was pretty good at judging people, said Adams "was always an honest man, often a wise one, but sometimes and in some things, absolutely out of his senses."

What did Franklin mean? Was John Adams crazy? No, he wasn't crazy, it just seems that sometimes he got carried away with his own ideas and forgot about reality. He had a hard time appreciating ideas that were different from his. Do you know anyone like that?

Sometimes when you study history it seems as if people in the past were all greater than people now. But they didn't look that way to the people who knew them. Ben knew John. He knew his good points and his weak ones. And Adams had plenty of both.

This was what people like Adams thought of France: a demon trying to bully America into going to war.

John Adams was brave and intelligent, and he loved his country. When he was young, and the country needed help breaking away from England, he was a strong leader and a fine thinker.

Then he went off to Europe, where he served his country

well as a diplomat in France, Holland, and England.

Perhaps he stayed too long in England. He grew fat and vain and peevish there. He grew to love English ceremony and English ways. He believed in representative government—what people called "republicanism"—but he didn't think much of democracy. Like Alexander Hamilton, he thought that the educated and the aristocratic should govern; he didn't trust the mass of people. Once, at a dinner party, Hamilton got angry at a Jeffersonian. Thomas Jefferson had faith that the ordinary people could govern themselves, but when Hamilton heard that idea he pounded the table with his fist and said what he thought. "Your people, sir," he said, "your people is a great beast!" And that was just what Hamilton, Adams, and the Federalists seemed to believe.

"Men are never good but through necessity," said John Adams, which means that people are good only if they have to be. Do you agree with that?

Many people do. And it may be true, but it wasn't true of either Hamilton or Adams. John Adams was a good man even when he didn't have to be. That means he always did what he thought was right—not necessarily what was popular. Remember, he was the lawyer who defended the British soldiers after the Boston Massacre. That certainly wasn't a popular thing to do.

John Adams was a complicated man. Thomas Jefferson, who was Adams's political opponent, wrote to a friend that John Adams was "so amiable, that I pronounce you will love him if ever you become acquainted with him." But Jefferson found out that Adams wasn't always amiable. Hamilton said Adams had a "temper."

Perhaps Adams was just too independent to be a good politician.

New DISPLAY of the UNITED STATES

JOHN ADAMS President of the United States

This engraving of President Adams shows the seals of all 16 states. What does "Millions for our Defense, not a cent for tribute" mean? (Hint: look up the XYZ Affair in an encyclopedia.)

**Amiable** means likable and easy to get along with.

Adams had a weakness for formal protocol. At receptions he wore velvet breeches and stood on a dais to greet guests.

John Adams said that "the affectionate participation and cheering encouragement" of his wife had been his "never-failing support."

Perhaps he had grown lazy by the time he became president. The old John Adams was different from the young John Adams. He was 61 when he became president, and he spent too much time at home in Quincy, Massachusetts, and too little time at the capital city. (Adams was away 385 days in four years as president; Washington was away 181 days in eight years.)

John Adams thought the best thing he did as president was to keep the United States out of war. He may have been right.

You see, France was fighting England. France had been America's best friend during the Revolutionary War, so the French thought the United States should side with them now against England. Some Americans agreed. Others remembered the old ties with England and wanted to back England. President Adams wouldn't let our nation take sides; he kept the United States neutral.

That made the French angry. They captured some American ships and took the sailors prisoner. That made a lot of Americans angry, especially Alexander Hamilton, who wanted to enter the war on England's side—even though England was also capturing American ships. John Adams had to fight Hamilton and other people in his own party. He did. He kept America out of war.

He tried to do something else and failed. He tried to stop some of the nasty political fighting between the Federalists and the Democratic-Republicans. He couldn't do that. Like Alexander Hamilton, John Adams was a Federalist. He believed in a strong central government.

As you know, the Democratic-Republicans wanted as little government as possible. They had faith that people could govern themselves. They believed in democracy. They called the Federalists "monarchists," which wasn't quite fair.

People in the two parties got very, very angry at each other. If children acted the way the country's leaders were acting, their parents would tell them to stop being silly, make up, and be friends. But each side was scared for the nation. Most of the Federalists really seemed to believe that if the Democratic-Republicans were elected the country was doomed. The Republicans believed that the Federalists had already messed everything up.

# 8 Alien and Sedition: AWFUL AND SORRY

**Adams's opponents often accused him of trying to make himself king of America. What do you think?**

Now the Federalists did something that was bad—dreadfully bad. In 1798, the Federalist Congress passed laws called Alien and Sedition acts, and President Adams signed them. (The Constitution says that the president must sign—or veto—all laws passed by Congress.) There were three Alien Acts—all were mean-spirited. One made it difficult for aliens—foreigners—to become United States citizens. Another said the president could throw anyone he wanted out of the United States, if he thought them dangerous. (President Adams never used that power.) The acts were aimed at the French, many of whom were fleeing France's revolutionary upheavals.

As you know, England and France were at war. The Federalists admired the English. (Thomas Jefferson and the Democratic-Republicans admired the French.) The Federalists acted as if the French were all villains. Religious prejudice was at work too. Most people in the young United States were Protestants. Most French people were Catholics. Unfortunately, some Americans wanted to keep Catholics out of the country. They supported the Alien Acts.

The Alien Acts were bad enough, but the Sedition Act may have been worse. It made it a crime to criticize the government.

Some people got arrested for doing just that. Some were newspaper editors. One was Ben Franklin's grandson. Another was Congressman Matthew Lyon of Vermont. Lyon had come to this country from Ireland, at age 15, as an indentured servant. He was one of the Green Mountain

*Do you know what happens to a law if the president vetoes it?*

## A Free Press?

Thomas Adams, the editor of the *Boston Independent-Chronicle*, attacked the Alien and Sedition acts and was thrown in jail. He continued to publish his newspaper from his jail cell. Across the top of the paper was printed: *A free press will maintain the majority of the people.* Next to it was the note: *This was originally written by John Adams...when the British excises [customs duties], stamp acts, land taxes, and arbitrary power threatened the people with poverty and destruction.*

43

**In 1789,** the United States got its first Roman Catholic bishop: John Carroll. (John Carroll was related to Charles Carroll of Maryland, who was the only Roman Catholic to sign the Declaration of Independence.) Some Americans feared the French revolution would send a wave of French Catholics to the United States. That was why a number of people supported the Alien Act. What do you think of that reasoning?

*Abridging* means cutting or limiting. A book that is abridged has been shortened.

**Remember** the Alien and Sedition acts and the resolutions when you read about the Civil War. The South will say the states have a right not to follow laws they think unconstitutional. They will be talking about laws that have to do with slavery. The South will quote the Virginia and Kentucky resolutions.

Boys who fought with Ethan Allen during the Revolutionary War. Lyon was an independent guy, wild, with a real temper. You probably would not have liked him. He was called "The Spitting Lyon" because he once spat in the face of Connecticut's Representative, Roger Griswold. Another time he had a fight—with fists and sticks—right on the floor of Congress.

Lyon attacked President Adams with words. It was in a Vermont newspaper, the *Rutland Gazette*. He said the president was trying to act like a king. He said that Adams should be sent "to a mad house." Well, because of the Sedition law, it was Lyon who got sent away—to jail!

All this happened in the country where people like Peter Zenger, Mary Goddard, and young John Adams had fought for free speech and a free press. Thomas Jefferson was talking about the Alien and Sedition acts when he wrote to a friend, "I know not which mortifies me most, that I should fear to write what I think, or my country bear such a state of things."

Article I of the Bill of Rights says: *Congress shall make no law respecting an establishment of religion, or prohibiting the free exercise therof; or abridging the freedom of speech, or of the press; or of the people peaceably to assemble, and to petition the Government for a redress of grievances.*

Congress and the president had done something the Constitution said they couldn't do. They were abridging freedom of speech and of the press. What could be done?

**The Virginia and Kentucky resolutions began an intense, sometimes violent debate over whether individual states could declare a law unconstitutional and then refuse to obey it.**

If Congress were to pass Alien and Sedition acts today, the Supreme Court would declare them unconstitutional. But the Supreme Court was just getting organized during Adams's term as president. The justices didn't even stay in one place; they rode around the country listening to cases. The court wasn't very strong.

No one was quite sure what to do. The legislatures of two states—Virginia and Kentucky—passed "resolves" (today they are called *resolutions*) declaring the Alien and Sedition laws unconstitutional. Jefferson wrote the Kentucky Resolutions; Madison wrote the Virginia Resolutions. They argued that if a state believed a law unconstitutional, it had the right to say so and not obey the law.

What Jefferson and Madison and most Americans wanted was to get rid of those awful Alien and Sedition laws. But just imagine if each state had the right to declare laws unconstitutional. Things would get very strange in this 50-state country.

Massachusetts replied to Virginia and Kentucky with its own resolutions. Massachusetts said that the states had agreed to the Constitution and were bound by that agreement. It was not up to the states, said Massachusetts, to say if a law was unconstitutional or not.

Is this getting complicated? Stick with it: it is very important. The Constitution was a great beginning, but there were things to be worked out. It took time to get the engine of government operating properly. It was not until 1803 that the Supreme Court first claimed the right to decide if a law is unconstitutional. That made a difference. It helped make our government work well.

The men who wrote the Constitution were afraid of power—political power. So they set up a government with three parts—the executive, legislative, and judicial branches—that were supposed to be equal partners to balance and check each other.

But at first the Supreme Court didn't seem to know what it was to do. It was no check or balance at all. The court was so weak it was even hard to get good people to serve as justices. Then President Adams made a brilliant choice. He appointed John Marshall as chief justice of the Supreme Court.

The term **checks and balances** is often used to describe our three-branch government. It means exactly what it says.

**When the** Alien and Sedition acts expired, they were not renewed.

Below, Matthew Lyon comes to blows in Congress.

FIFTH CONGRESS OF THE UNITED STATES:

Second Session.

Begun and held at the ci... ...lphia, in ...ate... f PENNSYLVAN... ...on Monday, the thir... ...and ...even hundred

An ACT in ...
...e United States.

BE it enacted by th...

# 9 Something Important: JUDICIAL REVIEW

**As a young** congressman, John Marshall voted against the Sedition Act and against his own Federalist Party. That took courage. He was chief justice of the Supreme Court for 34 years. He made the Supreme Court powerful and the judiciary an equal third branch of the government.

*Aristocracy* means government by a privileged class. Usually aristocrats are wealthy and powerful, but some people talk of an aristocracy of talent. In a letter to John Adams, Thomas Jefferson wrote, "I agree with you that there is a natural aristocracy among men. The grounds of this are virtue and talents."

John Adams, who appointed Marshall (above) chief justice, said: "My gift of John Marshall to the people of the United States was the proudest act of my life."

Virginia was ruled by an aristocracy. It was an aristocracy of mind as well as money. A poor boy with talent could make his way in Virginia. John Marshall was such a boy. Born in a log cabin on Virginia's frontier, he was the eldest of 15 children. His father was a farmer who helped George Washington survey some land. Washington became a family friend.

That friendship helped young John; but he would have succeeded anyway. He had two qualities that made that almost certain: he was friendly, cheerful, fun to be around— and he had a good brain and used it.

John Marshall had hardly any formal schooling: his parents were his teachers. They taught him well, and he studied and read on his own. When the Revolutionary War began his father enlisted. John went with him. He was popular in the army; someone with a merry nature was needed, especially during the terrible winter at Valley Forge. John was 22, tall, gangling, and good at athletics. He was known as "Silverheels," because he was a fast runner.

After the war, when his brothers and most of his friends headed west, John Marshall went the other way, to Williamsburg, where, for about six weeks, he attended lectures on the law by Thomas Jefferson's mentor George Wythe (say *with*). But, mostly, he learned law by studying on his own. When he opened a law office in Richmond he didn't have enough money to buy law books. It didn't matter; he had ability, ambition, and that easygoing, not-stuck-up nature.

Another lawyer described him thus:

> *A tall, slender figure, not graceful or imposing, but erect and steady. His hair is black, his eyes small and twinkling, his forehead rather low....His manners are plain yet dignified....His dress is very simple....I love his laugh—it is too hearty for an intriguer; and his good temper and unwearied patience are equally agreeable on the bench and in the study.*

John Marshall never seemed to take himself seriously. Someone who knew him said his clothes seemed "gotten from some antiquated slopshop." His dinner parties were famous for their good-natured, witty conversation.

"Simple as American life was, his habits were remarkable for modest plainness; and only the character of his mind, which seemed to have no flaw, made his influence irresistible upon all who were brought within its reach," wrote historian Henry Adams (who was John Adams's great-grandson).

In 1799, Marshall was elected to Congress as a member of the Federalist Party. The following year, President Adams named him secretary of state. The year after that he became chief justice of the United States Supreme Court. The Supreme Court met in the basement of the Capitol, because, although Pierre L'Enfant had planned a site for the Court, nothing had been built there yet.

But fancy rooms weren't any more important to John Marshall than fancy clothes. What he cared about was the way the United States was

**The author** of this description of John Marshall was Joseph Story, a famous lawyer and judge, writing in 1808.

John Marshall got little formal legal training. He learned the law by practicing it: taking on cases and arguing them in local courts such as this one (left). He argued only one Supreme Court case before he became chief justice—and he lost. Ironically, in that case he based his argument on the idea of states' rights; once he got to the Supreme Court, he spent much of his effort in striking down attempts to strengthen state power.

**Thomas Jefferson** didn't like the idea of judicial review; he thought it made the Court too powerful. A few legal scholars agree with him. Supreme Court justices are appointed, not elected. They hold office for life. (You can read the Constitution for details of this process.) Can you see problems if some not-very-good justices are appointed and they live for a long time?

*What does "unconstitutional" mean? If you don't know, see if you can figure it out, or ask for help.*

**Judicial** means "judges and court"; you know what **review** means. **Judicial review** means the review of laws by the courts.

governed. He believed that a strong government would help protect the rights of all the people. He tried to make the federal government stronger than the state governments. He tried to make the Supreme Court strongest of all. In 1803, in a very important Supreme Court case called *Marbury* v. *Madison*, Marshall said the Court could throw out any law passed by Congress if the Court thought that law was unconstitutional. "It is emphatically the province and duty of the judicial department to say what the law is," wrote Chief Justice Marshall in that very important case.

*Marbury* v. *Madison* began a process called "judicial review." It gave the Supreme Court the power to decide if a law passed by Congress meets the requirements of the Constitution.

But who really cares if a law is constitutional or unconstitutional, if Congress wants it? Well, imagine that tomorrow Congress passes a law saying you can't criticize the president. Suppose your mother does that and she goes to jail. That actually happens in some countries. In those countries people are even afraid to talk to their friends. It happened here in 1798 with the Sedition Act.

Judicial review protects all of us. It helps guarantee our freedoms. Judicial review made the Constitution stronger. It made the Supreme Court powerful. It made the Court a real check and balance to the two other government branches.

We Americans have always cared about our freedoms, especially the freedoms guaranteed in the Bill of Rights. John Marshall made sure those rights would be protected—even from Congress and the president.

"Nevertheless this great man nourished one weakness," wrote Henry Adams. "Pure in life, broad in mind, and the despair of bench and bar for the unswerving certainty of his legal method; almost idolized by those who stood nearest him, and loving warmly in return— this excellent and amiable man clung to one rooted prejudice: he detested Thomas Jefferson."

Does that surprise you? Well, here is another surprise: John Marshall and Thomas Jefferson were cousins. How did Thomas Jefferson feel about John Marshall? He loathed the man. Those two brilliant, remarkable Virginians couldn't stand each other. Each thought the other was not to be trusted. Their dispute was all about ideas. These were men who cared deeply about their ideas.

Now it just happened that John Marshall, with his relaxed, informal ways (in Richmond he sometimes went shopping with a market basket on his arm), was a supporter of causes that pleased the wealthy classes. While Thomas Jefferson, who had elegant taste and a French chef, was a champion of democracy and the common people. Which is just another of those paradoxes that keep turning up in history.

Marshall believed the purpose of the government was to protect "life, liberty, and property." (How did Jefferson define the purpose of government? Remember, it was "life, liberty, and…")

Neither Jefferson, nor Marshall, could quite believe that anyone who had conflicting ideas, and was also a deep thinker, could be honorable. Well, history shows that they were both honorable men. And, although they didn't know it, their ideas complemented each other. The nation needed them both.

John Marshall and his cousin Tom Jefferson did agree on one thing: the Alien and Sedition acts. Neither liked them. Most of the country didn't like them either. When John Adams ran for a second term he was defeated. Many people think it was because of his support of the Alien and Sedition acts. The great Federalist Party never achieved power again. In 1800 a Democratic-Republican was elected president. Guess who?

**John Marshall** got the idea for judicial review from George Wythe. Wythe talked about the need for judicial review when John Marshall was a student at the College of William and Mary.

The role of the Supreme Court (below) has been viewed differently at different times. Today it is generally seen as the national guardian of individual rights. How do you think it was seen in Marshall's time?

# 10 Meet Mr. Jefferson

Jefferson said, "Sound principles will not justify our taxing...our fellow-citizens to accumulate treasures for wars to happen we know not when."

Commemorative jugs were popular in the 1800s. This little pitcher had big words: "Hail Columbia, happy land...The rights of man shall be our boast, And Jefferson our favorite toast."

On a morning in March 1801, Thomas Jefferson sat down to breakfast at his usual seat at the end of a long table at Conrad and McMunn's boarding house in Washington, D.C., where he paid $15 a week for a room and three meals a day. The morning was cold and it was a special day, so someone offered him a seat near the fireplace. "No, thank you," said Jefferson, who would accept no favors. He meant to be a democratic president, a man of the people.

It was later that very day that he walked up the hill to the Capitol and was sworn in as the third president of the United States. Afterwards, in a quiet voice, he read his inaugural address.

Most of those listening were surprised by what he said. They expected something strong and startling from the man who had defied England with his great Declaration. Jefferson had spent the past years fighting the Federalists and Federalist ideas. Now he stood before them and said, "Let us unite with one heart and one mind. Every difference of opinion is not a difference of principle....We are all Republicans—we are all Federalists." It was an appeal for unity and good will. It set the tone for his presidency.

What was it about this man that made him so special? He was not a soldier. He was not a good orator. He was shy. He didn't pretend to be anything he was not.

Perhaps it was that Jefferson looked for the good in people. He appealed to the best instincts in his countrymen and women—and they knew it. He was, himself, a combination of the best the country had: his father was a farmer, his mother came from the Virginia planter aristocracy. From them he got a superb education and learned responsibility, good manners, and to be generous.

Jefferson wanted a government that would interfere as little as possible with people's lives. He cut taxes, reduced the size of the military, and balanced the budget.

People liked him. One group of Massachusetts Baptists liked him so much they decided to make a cheese for him. Their minister, John Leland, got the idea. Here is what they did, in the words of someone who was there:

> *Every man and woman who owned a cow was to give for this cheese all the milk yielded on a certain day—only no FEDERAL COW must contribute a drop. A huge cider press was fitted up to make it in, and on the appointed day the whole country turned out with pails and tubs of curd, the girls and women in their best gowns and ribbons, and the men in their Sunday coats and clean shirt-collars. The cheese was put to press with prayer and hymn singing and great solemnity. When it was well dried it weighed 1,600 pounds. It was placed on a sleigh, and Elder Leland drove with it all the way to Washington. It was a journey of three weeks. All the country had heard of the big cheese, and came out to look at it as the Elder drove along.*

That was one gift the democratic president couldn't refuse. It was called the "mammoth cheese," and everyone seemed to know about it. A poet wrote:

> *Some said 'twas Jefferson's intent*
> *To erect it as a monument.*

The cheese must have come in handy. It arrived in 1801 and was so enormous that in 1805 they were still serving it at presidential receptions.

President Jefferson's receptions were different from those of his predecessors. Mostly Jefferson gave small, informal dinners, but on New Year's Day and the Fourth of July the President's House was open to any citizen who wanted to meet the president. On ordinary mornings, if you had business to do with the government, you could stop by. Jefferson issued orders to his staff that all

---

**Thomas Jefferson**
PRESIDENT, 1801–1809

**If a nation expects to be ignorant and free, in a state of civilization, it expects what never was and never will be.**
—THOMAS JEFFERSON, 1816

***Predecessors*** are people who come before you. Who were Jefferson's presidential predecessors?

## The Big Cheese

Everyone was talking about the big cheese. Thomas Kennedy wrote a long and funny poem about it. The poem was published in *American Magazine* in 1802. Here is some of it:

Ye patriots now, of every state,
What wonders have you seen of late?
Great Leland rises to our view,
A patriot son, and reverend too.
His patriotism has been found
To weigh more than twelve hundred pound...
This Mammoth Cheese, a sight for all
True patriots, both great and small,
This priest attended day and night,
Lest Federal rats should get a bite.

Some accused Gallatin of trying to "stop the wheels of government," but he became a great treasury secretary—and was an expert on Indian tribes and their languages, too.

**Radical** means "from the root." It comes from *radix*, the Latin word for root. (So does the *radish*—which is a root vegetable.) A political radical wants to change things from the bottom up.

President Jefferson did not care for pomp; he rode horseback without a guard.

visitors—farmers and gold-braided diplomats—should receive the same courtesy. There was to be no favored treatment in a democracy. But when everyone was left to find his or her own seat at the president's table, there was so much pushing and shoving that, finally, seating charts had to be made.

When he had someplace to go, President Jefferson rode on horseback, without a guard—he had no use for the elegant presidential coach.

He chose his friend James Madison to be secretary of state and Swiss-born Albert Gallatin (GAL-uh-tin) as secretary of the treasury. As you know, many people say that Federalist Alexander Hamilton was the best secretary of the treasury ever. Well, others say the same of Republican Albert Gallatin. Gallatin said that the national debt was more dangerous to America than the risk of foreign aggression. The debt and the military budget had risen during John Adams's presidency. Gallatin reduced the military budget and cut the debt almost in half.

Jefferson believed he was involved in a revolution. He believed he was taking the nation back to the democratic spirit of 1776. Actually, he didn't change things as much as he thought he did. The country he led was in no danger of becoming a monarchy. George Washington and John Adams had enjoyed riding in a fancy coach, but they were republicans; they cherished this people's government.

And the Federalists—who had predicted terrible things from the man they screamingly described as a radical—were surprised. They had forgotten that Jefferson was a gracious country gentleman with fine taste and a belief in the goodness of men and women. The nation didn't fall apart under a Democratic-Republican administration.

But when Thomas Jefferson went shopping and bought a huge piece of land for the nation, some people thought it extravagant. Jefferson bought all the land that France claimed in North America. That land—the French called it Louisiana because their king was named Louis—went from the Mississippi River to the Rocky Mountains and maybe beyond. No one was sure how far it went. Jefferson spent $15 million on the Louisiana Purchase (which amounts to about four cents an acre). With that purchase he doubled the size of the country, and he did it peacefully. It was a great bargain, although at the time many thought it worthless and unneeded.

If President Jefferson hadn't bought that land, those of you who live in Missouri and Iowa and Nebraska, and a lot of other states, might all be speaking French today.

The Louisiana Purchase happened in 1803. That is a date to

**You may** wonder why France would sell all that land. Well, Napoleon, who was emperor of France, was fighting Britain. War is expensive. Napoleon needed the money.

Jefferson had wanted only to buy New Orleans, so American merchants could use the port duty-free. Then the French foreign minister Talleyrand (below, sitting) asked Secretary of State Monroe (left): "How much will you give for the whole of Louisiana?"

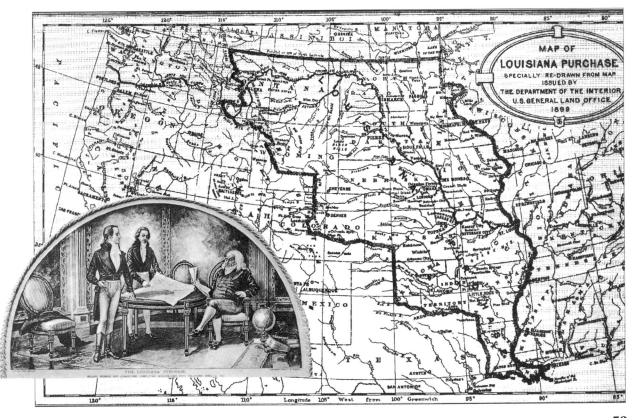

MAP OF
LOUISIANA PURCHASE
SPECIALLY RE-DRAWN FROM MAP
ISSUED BY
THE DEPARTMENT OF THE INTERIOR
U.S. GENERAL LAND OFFICE
1899

THE LOUISIANA PURCHASE

Hamilton (right) falls; at left, the dueling pistols that did the dreadful deed.

**Aaron Burr (above) was Jefferson's vice president because at the time that job was given to the man who came in second in the election. The growth of political parties made this practice unworkable, and it was changed by the 12th Amendment.**

**Alexander** Hamilton wrote a letter to John Adams and said that Aaron Burr was "unprincipled both as a public and private man…I feel it a religious duty to oppose his career."

remember. The Mississippi River was no longer controlled by a foreign power.

Once the United States purchased the Louisiana Territory, someone had to find out what it had bought. How big was the territory, what was it like, and where did it end? Jefferson sent an expedition to investigate.

Before I write about that exploring expedition I have something awful to tell you. It happened in 1804: Alexander Hamilton was killed in a duel with Vice President Aaron Burr.

Today dueling is against the law and the law is enforced. Then, people with arguments sometimes tried to shoot out their differences.

Aaron Burr was angry at Hamilton because Hamilton had supported Thomas Jefferson for president, instead of Burr. Alexander Hamilton supported Thomas Jefferson? Can that be true? Weren't they rivals? Yes, they were, but Hamilton was the kind of man who wanted to vote for the best qualified person. He knew Jefferson would make a better president than Burr. Besides, he couldn't stand Burr, who had once been a Federalist and then switched to the Democratic-Republican Party.

Hamilton was a man of rare talent and integrity who loved his country deeply. There are several different stories of that duel; people still argue about the details—but everyone agrees that Hamilton's death was a tragedy for the nation.

# 11 Meriwether and William—or Lewis and Clark

**Lewis (top) and Clark (bottom) took many gifts for the Indians they would meet. The expedition cost over $39,000.**

President Jefferson asked Meriwether Lewis to be his secretary. That didn't make much sense. Lewis was a terrible speller. Clearly, Jefferson had another reason for having Lewis around. He wanted to train him for an exploring mission. Thomas Jefferson was filled with curiosity about the West. He wanted to know about its land and its plants and animals; he wanted to know about the Indians who lived there. Are you ever curious about space and distant galaxies? The West was as unknown in 1803 as much of outer space is now.

Meriwether Lewis was born in Virginia's wooded piedmont. He became a captain in the Virginia militia. As a soldier he learned the ways of the Indians and how to survive in the wilderness. Lewis was a dreamer and a thinker, and, like Thomas Jefferson, a careful observer who loved the land and its birds and animals.

Just as we train astronauts today for voyages into the unexplored world, so, too, did President Jefferson see that Meriwether Lewis was trained in the scientific methods of the day. Lewis learned to gather seeds and identify

| **Lewis & Clark expedition** |
| 1804–1806 |

## Court Martial

The Commanding officers, Capts. M. Lewis & W. Clark constituted themselves a Court Martial [a military court] for the trial of such prisoners as are *Guilty* of *Capatal Crimes*, and under the rules and articles of *War* punishable by DEATH. *Alexander Willard* was brought foward Charged with *"Lying down and Sleeping on his post whilst a Sentinal, on the Night* of the 11th. Instant...." The Court after Duly Considering the evidence aduced, are of oppinion that the *Prisoner* Alexdr. Willard is guilty of every part of the Charge exhibited against him. it being a breach of the *rules* and articles of *War* (as well as tending to the probable destruction of the party) *do Sentience* him to receive *One hundred lashes, on his bear back, at four different times in equal proportion.* and Order that the punishment Commence this evening at Sunset, and Continue to be inflicted (by the Guard) every evening until Completed.                     Wm. Clark
                                        M. Lewis

**Rivers flow** from their source to their mouth. The *source* of a river is its beginning portion—usually a mountain stream. The *mouth* of a river is the place where it flows into another river, or into the ocean. Where is the mouth of the Missouri River?

Lewis learned the hard way how fierce a grizzly could be—and how fast a man could climb a tree.

Today's canoeists ride the rapids for a thrill; for the explorers they were just one more hazard.

Unable to travel during the winter, the expedition built a group of huts and called it Fort Mandan.

bones. Benjamin Rush—one of the most famous scientists in America—taught him how to preserve bird and animal specimens. Lewis's mother had medical skills; from her he learned to take care of himself and others.

William Clark was to be his partner in command. Together they prepared for a very difficult expedition. They chose men who were used to living in the wild. They trained them until they were tough and disciplined. When a man fell asleep on guard duty he was whipped. You can be sure he wouldn't fall asleep again.

Some people say it was the best organized exploration of all time. I'm not going to go that far, but it was pretty terrific. Meriwether Lewis and William Clark knew what they were about. They explored that big unknown land that the United States had just bought from France—the territory of Louisiana—and they even crossed the Oregon Country. They went all the way from the Mississippi River to the Pacific Ocean and back. It was dangerous country, with unexpectedly high mountains, difficult deserts, fierce animals, and wary Indians. They had prepared for danger, but they weren't quite prepared for the beauty: for the colors of wildflowers, the brilliance of sunsets on snowy mountain peaks, the sweet smell of prairie grass.

If only we could have been with them. They saw a world that would soon be gone forever. They saw birds and animals no white or black men had seen before—they saw woolly mountain goats and bighorn sheep and bright-plumed western woodpeckers. They dug up the bones of a 45-foot dinosaur. Wherever they went they took careful notes, made maps, wrote down vocabulary lists of Indian words, and collected samples of strange plants and animals. They added 200 species to the world's list of known plants. The Native Americans taught them to use some of those plants as medicines, some as foods.

If ever you need a partner for an adventure, try to find someone with abilities different from your own. Someone not like you, but whom you respect and enjoy. That was what made Lewis and Clark such a great team. They were not alike; their abilities complemented each other.

**The Mandan chief Sheheke went back East with Lewis. It took two tries and three years to get him home again, because of hostile tribes.**

Meriwether Lewis was a quiet, shy man. He liked being in the wilderness, away from civilization. He liked science and was a fine thinker, but sometimes he was moody.

William Clark was a happy, good-natured, talkative person. He was a redhead and the younger brother of George Rogers Clark, the Revolutionary War hero who has been called the "Washington of the West." William Clark, too, loved nature and the outdoors, but he also liked to be around people. Clark knew how to draw maps, and the actual maps he drew on that journey are now at Yale University. Like Lewis, he had been a soldier and was a Virginian. Clark was 34 that spring of 1804 when they headed out into the unknown West. Lewis was four years younger.

Here is part of the instructions Jefferson sent to Meriwether Lewis:

*The object of your mission is to explore the Missouri River...and communicate with the water of the Pacific Ocean. Beginning at the mouth of the Missouri, you will take observations of latitude and longitude at all remarkable points on the river....Your observations are to be taken with great pains and accuracy, to be entered distinctly and intelligibly for others as well as yourself to comprehend...these, as well as your other notes, should be made at leisure times and put into the care of the most trustworthy of your attendants...copies [should] be written on the paper of the birch [tree], as less liable to injury from the damp than common paper.*

*Other objects worthy of notice will be: the soil and face of the country, its growth and vegetable productions...the animals of the country generally, and especially those not known in the U.S.; the remains and accounts of any which may be deemed rare or extinct; the mineral productions of every kind...volcanic appearances; climate as characterized by the thermometer, by the proportion of rainy, cloudy, and clear days, by lightning, hail, snow, ice, by the access and recess of frost, by the winds prevailing at different seasons, the dates at which particular plants put forth or lose their flowers, or leaf, times of appearance of particular birds, reptiles, or insects.*

President Jefferson wanted to know all about the Native Americans who inhabited the land; he wanted Lewis and Clark to establish friendships with the Indians and prepare for trade with them. Treasury Secretary Albert Gallatin asked that they find out whether "that country

## A Sublimely Grand Specticle

*On June 13, 1805, Lewis beheld the Great Falls of the Missouri River. The spelling and grammar are his.*

**I had proceded about two miles...whin my ears were saluted with the agreeable sound of a fall of water and advancing a little further I saw the spray arrise above the plain like a collumn of smoke....I hurryed down the hill which was about 200 feet high and difficult of acces, to gaze on this sublimely grand specticle....the hight of the [right-hand] fall is the same of the other but the irregular and somewhat projecting rocks below receives the water in it's passage down and brakes it into a perfect white foam which assumes a thousand forms in a moment sometimes flying up in jets of sparkling foam to the hight of fifteen or twenty feet and are scarcely formed before large roling bodies of the same beaten and foaming water is thrown over and conceals them.**

*The story continues on page 60.*

BLACKFEET

WANAPAM

WALLA WALLA

CHINOOK

Fort
Clatsop

Mt. St. Helens

FLATHEAD

CLATSOP

Columbia River

Mt. Hood

NEZ PERCE

OREGON COUNTRY

SHOSHONI

Snake River

Pacific Ocean

MEXICO (Spanish)

ROCKY MOUNTAINS

The ROUTE of LEWIS and CLARK

CHINOOK

CLATSOP. WANAPAM. WALLA WALLA.

FLATHEAD

NEZ PERCE

SHOSHONI

HIDATSA

MANDAN

HIDATSA

Fort Mandan

CROW

MANDAN

ARIKARA

Lake Superior

Mississippi River

TETON SIOUX

YANKTON SIOUX

CHEYENNE    OMAHA

Missouri River

OTO

MISSOURI

St. Louis

LOUISIANA PURCHASE

The Corps:
LEWIS
CLARK
YORK
SACAJAWEA
"Seaman"—Lewis's
Newfoundland dog
New Orleans

ARIKARA    CHEYENNE    TETON SIOUX    YANKTON SIOUX    OTO    MISSOURI    OMAHA

On the way home, the team
split up. Lewis explored the
Marias River, a northern
tributary of the Missouri.

is susceptible of a large population," which is an old-fashioned way of
asking: Can lots of people live in the West?

Lewis and Clark went up the Missouri River on a 55-foot flatboat and
two narrow canoes. The boat held 21 bales of gifts for the Indians:
beads, ribbons, mirrors, cooking pots, and tools, as well as food and
supplies for the expedition. They moved slowly, mapping, exploring,
and hunting as they went. They wrote about their many adventures in
the daily journals they kept for the president. (You can find copies of
the journals of Lewis and Clark in most libraries.)

They expected to be able to get all the way to the West Coast by
boat, with perhaps a short portage when they reached mountains. They
knew that the source of the Missouri River would be found in mountain
streams. So they were prepared to climb mountains. On the other side
of the mountains they thought they would find new rivers leading to the
Pacific. Well, they were right—sort of. They found mountains, but not
the kind of mountains they expected. They were used to the time-worn
Appalachians. They weren't prepared for the awesome, towering Rocky
Mountains. They called them the "stone mountains." And the rivers
going west weren't where they thought they would find them. They had
to cross deserts to get to them.

That wasn't all. They were surprised by rattlesnakes, bears, and
mountain lions. But there were unexpected pleasures too. They feasted
on beaver tails, buffalo humps, and deer and elk steaks. They were
stunned by the endless herds of buffalo. They captured four black-and-
white magpies, put them in cages, and sent them back down the river
with other bird, animal, and plant specimens for President Jefferson.
Meriwether Lewis thought the caged birds were "butifull."

The Indians told them about grizzly bears; the bears were more dangerous than an armed warrior, they said. When an Indian wore a neck-
lace of bear claws, it was a sign of great bravery. Lewis and Clark didn't
quite believe the natives. They weren't scared; they had confidence in
their guns. So when they discovered bear tracks 11 inches long, they
were "anxious to meet with some of these bears."

It happened. Lewis and a companion spotted two grizzlies. They
both fired their rifles, hit their targets, and discovered that a single bullet will rarely kill a grizzly. One bear fled, but the other charged. Lewis
and his friend ran for their lives. They were lucky to survive. Their
muzzle-loading guns could not be fired again without reloading. The
bear, which turned out to be only a cub, was finally killed, but it took
several more shots. After that they boiled the bear's fat and turned it
into oil. Bear oil was a delicacy in the West; people liked it better than
lard, or even butter.

A man named York was an important member of the Lewis and Clark team. York was Clark's black slave. He was taller than six feet and an excellent swimmer, hunter, and trapper. The Indians were awed by York; most had never seen a black man before. Indian warriors often painted their bodies with charcoal. It was a mark of success in battle. So when they saw strong, charcoal-skinned York they thought him the mightiest of men. (York was freed when the expedition returned home. He headed back west and is said to have become chief of an Indian tribe.)

Even with all their training, Lewis and Clark might not have been successful—at least, they might not have gotten to the West Coast—if it hadn't been for a woman, an Indian woman named Sacajawea (sak-uh-juh-WE-uh). They met her when they built a camp and settled down for the winter in present-day North Dakota. Sacajawea was 16 and married to a Frenchman. She was about to have a baby; everyone was excited about that. When the baby was born she named him Jean Baptiste, for his father, but called him Pompey.

In the spring, when Lewis and Clark were ready to start out again, they hired Sacajawea's husband as a translator of Indian languages. Sacajawea came along with her baby strapped to her back. (Years later, Pompey became a well-known western guide.) But it was Sacajawea who turned out to be the helpful one; her no-good husband was lazy.

After following the Missouri River to its source, Lewis and Clark finally realized that there was no way to cross the continent by water. By now it was late summer. They had to get out of the mountains before winter; otherwise they might freeze or starve, and that would be the end of President Jefferson's exploration team.

They needed horses in order to continue. Would Indians sell them horses? Things didn't look good. Then some Shoshones came to their camp. As soon as Sacajawea saw the Shoshone leader she burst into tears. It was her brother, Ca-me-ah-wait. When she was a child, Sacajawea had been kidnapped from this very tribe. The white men had brought her home.

Ca-me-ah-wait was so happy he hugged Meriwether Lewis. "We wer all carressed and besmeared with their grease and paint," wrote poor-speller Lewis.

Lewis and Clark got their horses.

**Shoshone** means valley dweller.

Clark sketched this salmon in his journal. When they found salmon in the Columbia River, Clark knew they must be nearing the Pacific Ocean.

# 12 An Orator in a Red Jacket Speaks

The peace medal Washington gave Red Jacket bears the U.S. Great Seal. The eagle's olive branch and arrows stand for the power to make peace and war.

**The six** Iroquois nations are: the Cayuga, the Mohawk, the Oneida, the Onondaga, the Seneca, and the Tuscarora.

**The Iroquois** name for North America was "Turtle Island," which is why Sagoyewatha calls his country "this great island." What does he mean when he says, "from the rising to the setting sun"?

At the very time that Meriwether Lewis and William Clark were crossing the Rocky Mountains, a group of Indians was gathered on the other side of the continent, at Buffalo Creek, in the state of New York. They were Iroquois of the Seneca tribe and their leader was the powerful orator Sagoyewatha.

During the Revolutionary War (which all but the youngest remembered), the Iroquois had been divided. Some had fought on the American side, but others, like Sagoyewatha, had fought bravely for the British. Because Sagoyewatha wore a red English jacket in battle, the Americans called him Red Jacket.

At war's end Chief Red Jacket did not flee to Canada, as many Iroquois did. He stayed behind and was made to sign a treaty that gave much Indian land to the new nation.

Now the Iroquois were being asked to give up their religion. Some preachers had come from Boston to convert the Indians to Christianity. The Senecas listened politely to the ministers. Then Sagoyewatha/Red Jacket replied. His voice was strong, and mellow, and beguiling—fine speaking was a form of art and leadership to the Iroquois. This is what Sagoyewatha said to the Christian ministers:

*Friends and Brothers…this council fire was kindled by you. It was at your request that we came together at this time. We have listened with attention to what you have said.…*

*Brothers…listen to what we say. There was a time when our*

An Iroquois family on the hunt near Utica, New York. Red Jacket argued eloquently for the Indians' right to keep their way of life and practice their own religion.

*forefathers owned this great island…from the rising to the setting sun. The Great Spirit had made it for the use of Indians. He had created the buffalo, the deer, and other animals for food. He had made the bear and the beaver. Their skins served us for clothing. He had scattered them over the country and taught us how to take them. He had caused the earth to produce corn for bread. All this He had done for his red children because He loved them….*

*But an evil day came upon us. Your forefathers crossed the great water and landed on this island. Their numbers were small. They found friends and not enemies. They told us they had fled from their own country for fear of wicked men and had come here to enjoy their religion….We took pity on them …and they sat down amongst us….*

*The white people, brother, had now found our country. Tidings were carried back and more came amongst us. Yet we did not fear them. We took them to be friends. They called us brothers. We believed them….At length their numbers had greatly increased. They wanted more land; they wanted our country. Our eyes were opened and our minds became uneasy. Wars took place. Indians were hired to fight against Indians, and many of our people were destroyed. They also brought strong liquor amongst us. It was strong and powerful and has slain thousands.*

That was all history. It was the tale of the past.

**Thomas Jefferson** thought he had the Indians' interests in mind. First he wanted to protect them from gun-toting settlers; but then he hoped they would move west or learn to live like white people. (Do we think those the best choices today?) This is what he wrote: "Our system is to live in perpetual peace with the Indians.…by giving them effectual protection against wrongs from our own people.…They will in time either incorporate with us as citizens of the U.S. or remove beyond the Mississippi."

In youth Red Jacket was called Otetiani, meaning "prepared" or "ready." As a chief, he took the name Sagoyewatha—"He Causes Them to Be Awake." This may have referred to his skill as a speaker.

In George Washington's time, the Iroquois and the United States government were friendly. Here, Red Jacket (in front) and leaders of the Six Nations tour Philadelphia (see next page for more details). The Indians were treated like rulers of foreign nations (which they were). But the good relations did not last.

Now Sagoyewatha began to talk of the present and of the preacher's mission.

> *Brothers, our seats were once large and yours were small. You have now become a great people, and we have scarcely a place left to spread our blankets. You have got our country but are not satisfied; you want to force your religion upon us....*
>
> *Brothers, continue to listen. You say there is but one way to worship and serve the Great Spirit. If there is but one religion, why do you white people differ so much about it?...*
>
> *Brothers, we...also have a religion which was given to our forefathers and has been handed down to us, their children....*
>
> *Brothers, we do not wish to destroy your religion or take it from you. We only want to enjoy our own....*
>
> *Brothers, we have been told that you have been preaching to the white people in this place. These people are our neighbors. We are acquainted with them. We will wait a little while and see what effect your preaching has upon them. If we find it does them good, makes them honest and less disposed to cheat Indians, we will then consider again of what you have said.*

The missionaries loved their religion and wanted to share it. Were they wrong? Were they right? More than a century earlier, Roger

Williams had been able to love his religion and understand others, too. But there weren't many people like Roger Williams around—anywhere. Nor were there many people like Sir William Johnson. He had been able to live in two worlds: that of the Indians and that of the white settlers.

Most Americans didn't care about what happened to Red Jacket and his people. The Iroquois had lost most of their land and power. Things would get worse for them. They were going to be pushed west and then pushed again. There would be promises and treaties and they would all be broken. The new Americans wanted Indian land and they didn't know a fair way to share it. The Indians would lose most of their land. No one realized that in 1805, which was when Red Jacket spoke—though perhaps some understood that the Native Americans would not give up their land easily. Terrible Indian wars lay ahead.

If, in 1805, the Iroquois and the Christian missionaries could have looked into a crystal ball with a hundred-year vision, they would have seen the land west of the Appalachians filled with farms and cities. They would have seen new Americans living from coast to coast and the Native Americans squeezed onto reservations. That's what the land would be like in 1905. But if anyone had predicted that in 1805, he or she might have been thought crazy. The land west of the Mississippi seemed so vast that President Jefferson thought it would take thousands of years to fill.

**Jefferson said** there was "land enough for our descendants to the thousandth and thousandth generation." Was he right?

## From Feeble Plant to Mighty Tree

Red Jacket had been to the nation's capital (when it was in Philadelphia). That was back in 1793. He and 47 Native American leaders—Senecas, Onondagas, Oneidas, and Tuscaroras—came to meet the president and to negotiate treaties to hold on to their lands. The Speaker of the House of Representatives took them on a tour of Philadelphia, and President Washington gave each a silver peace medal. The Indians gave presents of their own and demonstrated some of their dances. Then Red Jacket spoke. Even among the eloquent sachems, he was known for his gifted tongue.

*We first knew you a feeble plant which wanted a little earth whereon to grow. We gave it to you: and afterward, when we could have trod you under our feet, we watered and protected you: and now you have grown to be a mighty tree, whose top reaches the clouds, and whose branches overspread the whole land, whilst we, who were the tall pine of the forest, have become a feeble plant and need your protection.*

The reverse side of the peace medal: Washington presents a peace pipe.

# 13 The Great TEKAMTHI, Also Called TECUMSEH

"We gave them forest-clad mountains, valleys full of game," said Tecumseh. "In return they gave us...rum and trinkets and a grave."

It was the ninth of March in the year 1768 when a great meteor—a shooting star—flamed across the sky with a light bright as fireworks.

In the English colonies those who saw it remarked about the beauty of the heavens; some said it was an omen, a sign of changes to come.

The Indians, too, saw it as an omen. The Shawnee tribes told stories of the stars: this one, the shooting star, was a great spirit called "the Panther." Each night it passed somewhere over the earth, heading for a deep hole and sleep.

Pucksinwah, the Shawnee chief, was awed by the star. He had never seen one of such brilliance. At the very moment the star burst over his head he heard the cry of a new baby. It was his son, born under a shooting star. Pucksinwah knew it was a good sign. He named the boy "The Panther Passing Across." In the Shawnee language that was *Tek-am-thi*.

The boy did not disappoint. Before he was 10 it was known that he would be a leader. He could run faster than the others, he could shoot an arrow straighter, he could re-member more, and he didn't brag.

At 11 he had a new friend: a white boy was adopted into the tribe. It was not unusual to have whites become Indians. Some were captured in raids, some were orphans who need-ed care, some chose to become Indians. (The great Shawnee

chief Blue Jacket, though he tried to hide it, had once been white.)

Tekamthi's new friend taught him to speak English, and, since Tekamthi had a quick mind, he learned easily. He wanted to learn more, so the boy taught him to read and write. That was unusual. Many Indians learned to speak English; few could write it.

Tekamthi learned other things: to love the land and its animals; to know the plants that heal and those that harm. He learned to hunt and was soon the best in his tribe. And he learned of the Great Spirit who ruled the earth and skies, and Tekamthi believed.

He met the white men who were coming into his land—the land of Kentucky and Ohio. He respected the brave men like Daniel Boone, but others he grew to hate. For they killed his father and took his land and made promises they did not keep.

Always the white men told the Indians that if they just moved once more they would be secure. If they just signed a treaty, they would have land and would not have to move again. And some of the Indians believed them. Tekamthi did not.

He wanted the white men to go back—over the mountains—and

The Prophet, Tenskwatawa— "The Open Door."

Richard Johnson (on horseback) claimed to have killed Tecumseh (far right). He called himself "Old Tecumseh" to further his political career.

**William Henry Harrison, who was a young man when he fought against the Prophet and his people, became our ninth president in 1841.**

**The Shakers**, a Christian sect that used shaking dances in worship services, called themselves the United Society of Believers in Christ's Second Appearing.

leave the hunting lands of the West for the Indians. He would make the white men go. He would do it by uniting the Indian tribes. One strand of hair, he said, is easily snapped. But a thick braid is almost impossible to tear.

He would braid the tribes into a mighty league. His brother would help him.

Tekamthi was muscular and well-built, and his face was so handsome that men and women remarked about it. When he danced the part of a warrior, it was with such strength that everyone forgot that he had a wounded leg, which gave him a limp.

Tekamthi's brother, Tenskwatawa, was small and homely, but he was a shaman (SHAY-mun)—a religious leader. He was called "The Prophet," and was renowned for his wisdom. Together the brothers made an awesome team.

They told the Indians to stop drinking the white men's liquor—that it only made them weak. And the tribesmen stopped. They told them to go back to Indian ways and to be proud of their heritage. And the Indians did that too.

Tekamthi wished to lead his people in the ways of goodness. He wished to follow the best of the Great Spirit's teachings. His people believed in him.

Tekamthi traveled far to reach other tribes: he went to the land the white men called New York and then to lands west of the great Mississippi. Everywhere he gathered followers, although when he went South, to visit the powerful Cherokees, their chiefs would not join with him. They liked the ways of the white people. Tekamthi said he did not want to fight the white men, he wanted to share the land, but if the whites would not share, he would fight.

The whites called him Tecumseh and knew he was powerful. Some white Protestant ministers—members of the Shakers—came to listen to Tecumseh and his brother the Prophet. They were surprised by what they heard and mightily impressed. They wrote a report in May 1807. This is what they wrote:

> *Our feelings were like Jacob's when he cried out "surely the Lord is in this place, and I knew it not...." Although these poor Shawnees have had no particular instruction but what they received from the outpouring of the Spirit, yet in point of real light and understanding, as well as behavior, they shame the Christian world.*

William Henry Harrison, who had been appointed governor of the Indiana Territory, was worried about Tecumseh. Most of the Indiana Territory, by treaty, was supposed to be Indian land. But white settlers were moving in. Harrison was afraid that Tecumseh was too powerful, that the Indians would endanger the white settlers.

William Henry Harrison—tall, slim, and soldierly—had been trained to be a leader. He was the son of Benjamin Harrison, who signed the Declaration of Independence and was governor of Virginia. Young Harrison wished to prove himself. So, in 1811, when he knew Tecumseh was far away visiting tribes in Alabama, Harrison marched to the Shawnee camp on the banks of the Tippecanoe River. The Prophet was in charge. He knew nothing of military leadership; he thought his belief in the Great Spirit would be enough. The Prophet told his followers that God would make the bullets bounce off their chests. He may have believed that. When Harrison and his army were just a mile away from the Indian village, the Prophet and his men attacked.

**After the** battle in which Tecumseh was killed (no one knows by whom), his body was carried from the field and buried secretly in a grave that has never been discovered.

JOURNEYS of TECUMSEH

Trapped by Harrison near Lake Erie when the British retreated, Tecumseh fell at the Battle of the Thames.

**The battle at Tippecanoe ended any hope for the creation of a united Indian nation that could forge a peace between the Native Americans and the whites.**

Bullets did not bounce off their chests. Indians died. Actually, two of Harrison's soldiers died for every Indian killed. But the Shawnee village was destroyed and the Shawnee hearts went with it.

Harrison claimed a great victory at Tippecanoe. It made him a national hero. His nickname became "Old Tippecanoe."

The tribes would no longer unite behind Tecumseh. No leader was ever able to braid them together. In 1813, Tecumseh fought with the British against the Americans and was killed in battle. White men moved into the Indiana Territory. The treaties with the Indians were forgotten.

# 14 Osceola

Osceola's mother had a second husband, a white man named Powell. Osceola (above) was sometimes mistakenly called a half-breed with the Powell surname.

The Indian boy was six, or perhaps seven, when the great Tecumseh came to his village on the Tallapoosa River in Alabama. Tecumseh had left his tribespeople by the waters of Tippecanoe; he had traveled south to meet with the Cherokees, the Creeks, the Choctaws, the Chickasaws, and the other peoples of the great Indian confederation that stretched across the lands the white men called Mississippi and Georgia and Alabama and Carolina. Tecumseh and his men performed feats of strength and daring. The boy, a Creek of the Tallasee tribe, must have been mightily impressed. But did he understand the seriousness of their mission?

Tecumseh had come to persuade the tribes to forget their rivalries. He preached a message of strength in unity. Tecumseh's mother was a Creek. His kinsmen listened politely, but they did not hear.

There were two factions among them: some, who lived and farmed as the whites did, were called White Sticks. The others, who still maintained their Indian ways, were Red Sticks. The White Sticks wanted to cooperate with the Americans; the Red Sticks wanted to drive them away. When Tecumseh died, there was no leader left who even attempted to hold the Indians together. Tecumseh was right: they would lose their only chance to change American history. The white men would use them. In the war that was coming—the War of 1812—the White Sticks would fight with the Americans against the Red Sticks and the British. It would be an Indian civil war.

It all came to a climax in March 1814, when General Andrew

**Between 1789** and 1825, the Creeks, Cherokees, Choctaws, and Chickasaws made 30 treaties with the United States. With all of these treaties the Indians surrendered lands or agreed to new boundaries.

**When Tecumseh** departed the Creeks' country to continue his mission, he left a relative to instruct the Creeks in the religious Dance of the Lakes—and in the use of magic red war clubs that would protect the Indians and drive the white men into quagmires. Thus Tecumseh's followers became known as Red Sticks, after their clubs.

A *faction* is a group of people who share views—especially political views—like a political party.

In 1790 the artist John Trumbull drew these portraits of two Creek chiefs who visited President Washington in New York to discuss giving up some Indian lands. Trumbull said that the chiefs had "a dignity of manner...worthy of a Roman senator."

*Opulent* means lavishly rich and abundant.

Jackson and the White Sticks fought a big battle at the Horseshoe Bend of the Tallapoosa River. They beat the Red Sticks. Although many didn't understand it at the time, it was a disaster for both groups of Indians. Never again were the Creeks powerful.

Soon after the battle at Horseshoe Bend, the Americans thanked their White Stick allies by forcing them to sign a treaty handing over most of their land. The Red Sticks had worse problems—they now had no land at all. They hid in the woods. The Creek boy, his mother, and what was left of the Tallasee tribe headed south. They crossed the border into Spanish Florida.

The Indians who lived in Florida were called Seminoles. They were a mixture. Many had been Creek; a few were of the old, mostly extinct Florida tribes; some were wanderers and outcasts; some were blacks who had been slaves. The word *Seminole* was said to have come from an Indian word meaning "runaway," although some said it was a mispronunciation of a Spanish word meaning "wild ones."

The Creek boy and his tribespeople were welcomed by the Seminoles. They settled in fertile north Florida. In that opulent land they could feast on fish cakes, corn bread, bear ribs, alligator steak, turkey legs, and delicate rattlesnake meat.

The boy grew to love his new home. He and his people kept many of their Creek ways. The women farmed; the men hunted. Because the climate was warm, their houses were simple roofed platforms on stilts that kept them off the moist ground. The houses were built around a central square that was used for ceremonies and tribal dances.

The boy was of average height, with small hands and feet and a sensitive, almost pretty face. It was when he played the Indian ball game the French called *lacrosse* that his energy and determination became apparent. He liked to win, and he usually did. When he was 15 he was old enough to take part in the sacred ceremonies that introduced him to manhood. He fasted and prayed and drank of the bitter black juice that marked his coming of age. Now he received his adult name. He was Asi-yo-ho-lo. The white men who heard it said *Osceola*.

Like Tecumseh, he stood out among the others. He seemed braver and more truthful. But he carried with him anger at White Stick Creeks and at white men. You will hear more about Osceola.

The Red Stick Indians fought troops under Andrew Jackson (above) at Horseshoe Bend with incredible bravery. By the end only a handful survived.

## Take Me Out to the Ball Game

*George Catlin was a lawyer, an artist (he painted the portrait of Osceola in this chapter), and a student of the ways of life and customs of North American Indians. Here he illustrates and describes a Choctaw ball game, the same sport of lacrosse that Osceola played.*

Each party had their goal made with two upright posts, about 25 feet high and six feet apart, set firm in the ground, with a pole across at the top. These goals were about 40 or 50 rods apart; and at a point just half way between, was another small stake, driven down, where the ball was to be thrown up at the firing of a gun, to be struggled for by the players...who were some 600 or 700 in numbers, and were mutually endeavoring to catch the ball in their sticks, and throw it home and between their respective stakes....every mode is used...to oppose the progress of the foremost, who is likely to get the ball; and these obstructions often meet desperate individual resistance, which terminates in a violent scuffle, and sometimes in fisticuffs....For each time that the ball was passed between the stakes of either party, one was counted for their game... and so on until the successful party arrived to 100, which was the limit of the game, and accomplished at an hour's sun.

# 15 The Revolutionary War Part II, or the War of 1812

**James Madison**
PRESIDENT, 1809–1817

## Poor Thinking

Captain Oliver Hazard Perry was one of the best American sailors of the War of 1812. But when Commodore Chauncey sent him some black men as reinforcements before the battle of Lake Erie, Perry complained. He thought the blacks would be no good. Chauncey replied:

*I have yet to learn that the color of the skin...can affect a man's qualifications or usefulness. I have 50 blacks on board this ship and many of them are my best men: and these people you call soldiers have been to sea from two to 17 years: and I presume you will find them as good and useful as any men on board your vessel.*

Lady Liberty—and the victims' heads—look on as Madame Guillotine does her work.

George Washington warned Americans to stay out of foreign quarrels. Unfortunately, England and France wouldn't let us do that. It would have been nice if they had left our infant nation alone. This country needed a few years to grow up without having old friends pick on us. But that wasn't to be. And so we went to war. This is how it came about.

As you know, soon after our revolution, France had a revolution of its own. At first, everyone thought the French Revolution would be like the one in America—but it got out of hand. The winners of the revolution began chopping off the heads of the losers—really—with a head-chopping machine called a guillotine.

Then Napoleon came to power. He tried to conquer most of Europe, and England declared war on France.

Americans had a hard time trying to figure out what was going on across the ocean. News traveled slowly. We tried to stay out of the European war; we tried to stay neutral, but neither England nor France respected nations that were neutral.

Both countries captured American ships and took American sailors as prisoners. The British sea captains forced the American sailors to work on British ships. Naturally, that made people in this country angry.

There was something else that made Americans angry with England. England had never cleared out of the forts it held in the territories west of the Appalachian Mountains. (England was supposed to leave after the Revolutionary War.) Many English men and women seemed to believe that someday America would again be part of the British Empire.

Capturing ships and holding territory—those are two pretty good reasons for fighting. There was a third reason that wasn't so good. Many Americans wanted Indian land. But the English in their western forts had become friends and protectors of the Indians.

Americans were divided. Most New Englanders did not want to get involved in another war with England. (England controlled the seas, and New England depended on its sea trade.) Most Westerners, and some Southerners, wanted to go to war. Scholarly James Madison, who was 63 and president, was cautious. But in Congress and across the land, a new generation of leaders was taking the place of the men who had fought in the Revolution. These new leaders had been youngsters when the Constitution was written. They didn't have memories of the old days when there were good feelings toward England; they didn't

**James Madison**
was Thomas Jefferson's good friend. Jefferson's home, Monticello, was not far from Madison's home, Montpelier. Madison wrote the Bill of Rights and guided the writing of the Constitution.

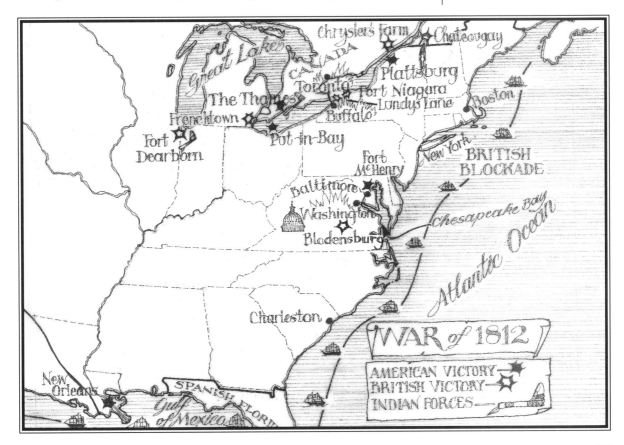

remember the long, hard battles of the Revolutionary War. All they had heard were stories of battlefield valor and victory. They were eager to fight. Henry Clay, from Kentucky, was one of the new leaders; Andrew Jackson, of Tennessee, was another. Those who wanted to go to war were called "war hawks." They finally convinced President Madison that the nation's honor was at stake. In 1812 war was declared against Great Britain.

The War Hawks thought they could march to Canada and take all that land from England. Henry Clay boasted that the Kentucky militia could do it all by itself. When it came to the sea, however, he thought differently. So did most Americans. They believed the famous English navy would easily outfight them. Were they surprised! Canada didn't fall to the American forces—the Americans fought poorly in Canada—but the little American navy won some surprising victories against the big British navy, and no one in the world expected that.

But the War of 1812 (which really should be called the War of 1812 to 1815, because that was how long it lasted) started slowly. Then the war in Europe—the one between England and France—ended. Napoleon lost. That was terrible for the United States. It meant the English could now send more soldiers to America. They did. In 1814 they sailed an army into Chesapeake Bay (check the map on page 75), landed redcoats, whipped American troops at Bladensburg (five miles from Washington, D.C.), and then marched on, heading for the nation's capital.

By that time there were hardly any American soldiers in the city— they'd all fled—and those residents who were still around were keeping quiet about it behind their shuttered windows. The British commander, Major General Robert Ross, had orders to "destroy and lay waste" the towns he captured. Ross, however, was prepared to negotiate with Washington's leaders: if they paid ransom he might spare the city. But there weren't any leaders to be found.

So the general sent his soldiers to burn Congress's home—the Capitol—but it was so well built they had a hard time of it. They piled up chairs and books and drapes and torched them. Finally, much of the building went. Then they headed toward the President's House.

James Madison wasn't at home. (He had been at Bladensburg; now he was off somewhere consulting with his generals. He was courageous, but when it came to military leadership he was a dud.) It was his wife, Dolley, who was the indomitable member of the family. She was in the President's House getting ready to have dinner with friends. Dolley was advised to flee, but, before she did, the First Lady took important papers, silverware, and red velvet drapes, and had them loaded onto a wagon. Then she waited while a portrait of George Washington was

When the Madisons returned to the capital, Dolley said, "We shall rebuild Washington. The enemy cannot frighten a free people."

**During the** War of 1812 the British burned the Library of Congress and all its books. In order to start a new library, the nation bought 6,487 books from Thomas Jefferson for $23,950 (which was much less than they had cost Jefferson).

torn from the wall (it took several men to do it). She left instructions to have the painting packed up and sent to safety; she got out just before the British arrived. The British toasted the king with the president's Madeira wine and ate Dolley's dinner—the wine was still chilled and the meats laid out handsomely. After dinner the British set a bonfire inside the President's House.

From there they carried their torches to the Treasury and were disappointed to find no money inside. They burned that building and the Library of Congress, too. Dr. William Thornton, the architect of the Capitol (remember, he won the contest), was brave enough to stand outside the Patent Office and convince the soldiers that they would be destroying inventive science if they burned that building. So they spared it, but, the next day, a storm erupted with wind and rain so intense that the roof was blown off the Patent Office and some

**Many blamed Madison (fourth from left, carrying papers) for the war and the destruction of Washington. This caricature suggests that Madison was too friendly with Napoleon and so had attracted British fury.**

77

**Was the** War of 1812 worth it? Well, it made the British realize they had really lost their colonies. It made them respect the United States. It made us Americans realize we could not have Canada. It made our nation grow up. It made Americans feel proud. After all, we had fought off the most powerful nation in the world—twice.

inventions with it.

The British soldiers were now eager to march on—to Baltimore. They hated that place. With its 45,000 inhabitants, it was the fourth-largest city in the States. But it was Baltimore's private sailing ships that made them angry. Privateers from Baltimore had been capturing and sinking British ships. No mercy was to be shown at Baltimore. Word went up and down the coast that Baltimore was expected to fall, and, after Baltimore, who knew where the British would head next. Defeatism and gloom descended over much of the country. Citizens from Alexandria and Georgetown rushed to surrender to the British—without even being asked. Annapolis prepared to do the same. But the British fleet sailed right by that town—they were after the bigger city up the river. They called it "Mobtown."

They didn't know that something was happening at Baltimore that was changing the mood of despair. Major General Samuel Smith, who

This aquatint of the battle of Fort McHenry shows very clearly the "bombs bursting in air"—while a farmer and his cow in the foreground carry on calmly as if nothing were happening.

had leadership genes in every one of his cells, had taken charge of the city. Smith wasn't giving anyone time to think of surrender. He put Baltimore's citizens to work. It wasn't easy for him to get the rich merchants to agree to sink their own ships in the harbor to make a barrier to keep out the English, but Smith convinced them that it was better than losing everything if the city fell. Trenches were dug, cannons put in place, and the citizen militia—which had performed miserably at Bladensburg—was drilled and drilled again. A letter printed in the *Evening Post* said, "White and black are all at work together. You'll see a master and his slave digging side by side. There is no distinction, whatsoever."

Baltimore harbor was guarded by big, star-shaped Fort McHenry. Major George Armistead commanded that fort. Early in the war, Armistead went to Mary Pickersgill, a well-known Baltimore flagmaker, and ordered an American flag "so large the British will have no difficulty seeing it at a distance." She and her 13-year-old daughter sewed stripes of red-and-white English wool bunting and added cotton stars. The huge flag was 42 feet by 30 feet, had 15 stripes and 15 stars, and cost $405.90. Armistead raised it over the fort.

The English ships couldn't miss it when they sailed into the harbor. They were soon approached by a small ship of truce. Francis Scott Key, a Washington lawyer who had fought at Bladensburg, had been sent to get William Beanes, an elderly, lively, and much-loved doctor taken as a hostage by the British. Key brought letters from British prisoners of war stating that they were being well treated—that got Dr. Beanes released. But the truce ship was not allowed to return to the American side. Francis Scott Key and Dr. Beanes had to wait with the British until the fighting was over.

The battle began on a Sunday. "May the Lord bless King George, convert him, and take him to heaven, as we want no more of him," said the Reverend John Gruber at Baltimore's Light Street Methodist Church. Then his congregation rushed out and prepared for whatever was ahead of them. At about the same time, the British admiral, Sir George Cockburn (pronounced CO-burn), was changing his plans. He could see that the ships sunk in the harbor prevented him from sailing right up to the city. He would have to capture Fort McHenry if he wanted Baltimore.

That night, rockets and bombs lit the starless sky like fireworks. Some of the British ships' cannons had a range of two miles—nothing in the batteries at Fort McHenry could match that. Early in the morning a British force set out to do battle. Then something happened that made a big difference: the British commander, General Robert Ross, was killed. The English soldiers were now rudderless.

**After the** war, the President's House was painted white to cover the scorch marks left by the British. That's when everyone began calling it the "White House," which is what it is called today.

Samuel Smith had distinguished himself in the Revolutionary War 35 years before.

**Mrs. Pickersgill's** flag (which may be seen today at the National Museum of American History in Washington, D.C.) had 15 stripes and 15 stars. At first, stars and stripes were added for each new state. Then the flag got to be an awkward size. Finally, the flag had 13 stripes for the original colonies, and stars for each state. (How many stars are there now?)

## The Last Battle

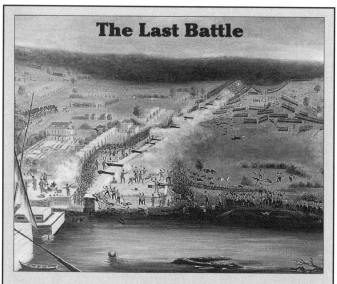

**D**o you know what the word *irony* means? Hint: It has nothing to do with metal and everything to do with things not being as they seem to be. There was a big irony in the War of 1812. It was about a battle fought at New Orleans on January 8, 1815: big battle, big American victory, big irony.

The Americans lost 13 men at New Orleans, and about 60 were wounded. More than 2,000 British soldiers were killed, wounded, or missing. The hero of that astounding battle was little known. He was the same general who had fought Osceola's people at Horseshoe Bend. His name was Andrew Jackson. So what was the irony?

A peace treaty had been signed a few days earlier in Europe, but the news hadn't reached New Orleans. No one knew the war was over. All those men—on both sides—died fighting a war that was already history.

The war made Andrew Jackson a national hero. After the battle of New Orleans, Americans sang this song:

> *But Jackson he was wide awake*
> *And wasn't scar'd at trifles*
> *For well he knew what aim we take,*
> *With our Kentucky rifles,*
> *So he led us down to Cypress Swamp,*
> *The ground was low and Mucky,*
> *There stood John Bull in martial pomp*
> *But here was old Kentucky.*

John Bull is the symbol of England; old Kentucky was the Americans —called that because of their guns (known as Kentucky rifles).

While British ships were fighting and shooting shells on the Fort McHenry side of the Baltimore peninsula, British infantry—foot soldiers—were marching up the other side. They were supposed to coordinate an attack. But there was a communications foul-up. The soldiers turned back, and the navy didn't know it. They fought to protect the soldiers when they weren't where they were supposed to be.

The bombing of Fort McHenry continued through the day and on into the second night; it went on for 25 hours. Francis Scott Key, on the truce ship, watched the "rockets' red glare" and listened to "bombs bursting in air." And then it stopped. It was very dark, and very quiet, and it was raining so hard that no one could tell what had happened. Had the fort been captured? No one knew, until, finally, "by the dawn's early light," Key saw Mrs. Pickersgill's "broad stripes and bright stars," and realized—the fort had held!

Now it was the British who retreated. Francis Scott Key found words spinning in his head. He wrote them down; he thought they could be sung to an old British drinking song called "To Anacreon in Heaven." (Key had set words to that song once before. Most people knew the tune.) Some friends published his poem in Baltimore, calling it "The Defense of Fort McHenry." Soon it was republished in town after town. The heroic action in Baltimore inspired the young nation. If Baltimore had burned, where might the British have marched next? Philadelphia had readied its militia. Now, everywhere, there was relief, jubilation, and celebration. People were soon singing Key's words, and someone gave them a new name. It was "The Star-Spangled Banner."

# Our National Anthem:
# The Star-Spangled Banner

In the first stanza, old Dr. Beanes is speaking. He is asking Francis Scott Key some questions. He uses a few unusual words: *ramparts* are high walls that surround a fort. *Perilous* means dangerous. *O'er* is the poet's way of saying "over."

The reason Francis Scott Key (left) and Dr. Beanes had to stay aboard the truce ship: they had overheard the British plan of attack, and so were held to stop them taking the news ashore.

*Oh! say can you see, by the dawn's early light,*
*What so proudly we hailed at the twilight's last*
*gleaming?*
*Whose broad stripes and bright stars, through the*
*perilous fight,*
*O'er the ramparts we watched were so gallantly*
*streaming.*
*And the rockets' red glare, the bombs bursting in air,*
*Gave proof through the night that our flag was still*
*there.*
*Oh! say, does that star-spangled banner yet wave*
*O'er the land of the free and the home of the brave?*

Does the flag still wave? Has the fort held? In the second stanza, poet Key answers the questions. More words: *foe* is enemy; *towering steep* are other words for those steep walls—the ramparts.

*On the shore, dimly seen through the mist of the*
*deep,*
*Where the foe's haughty host in dread silence reposes.*
*What is that which the breeze, o'er the towering steep,*
*As it fitfully blows, half conceals, half discloses?*
*Now it catches the gleam of the morning's first beam,*
*In full glory reflected, now shines on the stream:*
*'Tis the star-spangled banner, Oh! long may it wave*
*O'er the land of the free and the home of the brave!*

We're skipping the third stanza—which is not often sung. The fourth stanza is about hope for America's future. It is usually sung a bit more slowly than the others.

*Oh! thus be it ever, when freemen shall stand*
*Between their loved homes and the war's desolation.*
*Blest with victory and peace, may the Heav'n-rescued*
*land*
*Praise the Power that hath made and preserved us a*
*nation.*
*Then conquer we must, for our cause it is just.*
*And this be our motto—"In God is our trust."*
*And the star-spangled banner in triumph doth wave*
*O'er the land of the free and the home of the brave.*

81

# 16 That Good President Monroe

Our country is like a new house, said Monroe. "We lack many things, but we possess the most precious of all—liberty!"

**At Monroe's** swearing-in for his second term as president the band played "Hail to the Chief"—the first time that song was played for a president. "Hail to the Chief" was the work of two Scotsmen: the composer, James Sanderson, wrote the tune to accompany lines from a poem by the famous writer Sir Walter Scott.

James Monroe reminded some people of George Washington. He was another tall, courtly Virginian, and so honest that Thomas Jefferson said you could turn his soul inside out and "there would not be a spot on it."

When Monroe became our fifth president, in March 1817, he was already being called the "last of the Revolutionary farmers." Washington, Adams, Jefferson, and Madison were all unusual farmers: each loved and understood the land, each revered book learning, and each believed it was his duty to serve his country. James Monroe was the same kind of man.

People liked him. It made them feel good to have a president who was handsome and kindly, who had fought bravely as a soldier in the Revolutionary War, who had studied law with Thomas Jefferson, and who had served in the Virginia General Assembly when the Constitution was ratified.

Monroe wore knee pants and silver buckles on his shoes, even though those styles were old-fashioned by the time he was president. He was tall and gangly, with wide-set gray eyes and a large nose. He had been raised in a privileged home where he was trained to be a leader and to help others.

Some important things happened during his presidency. One important thing was that the United States got Florida from Spain.

The White House was painted white in 1817, the year Monroe became president. John Quincy Adams, who succeeded Monroe, often slipped down to the nearby Potomac River for a swim.

(That is especially important if you happen to live in Florida.) Spain—the country that was so mighty in the time of Columbus—had become weak.

Slaves had been fleeing to Spanish-held Florida, where they lived in all-black villages or in bi-racial Seminole communities. Suppose you were a slave, and willing to risk the dangers of running away. If you lived in Virginia, or Maryland, or Kentucky, you would probably head north and try to make it to Canada. But if you lived in South Carolina, or Georgia, or Alabama, you wouldn't have much chance of escape if you went north. If you headed the other way, and were lucky, you might just get to Florida. There the Seminole Indians would protect you. You can see why the Southern slave owners hated the Seminoles. And they were angry with the Spaniards for not making the Seminoles return their runaways.

In 1817, Secretary of War John C. Calhoun sent General Andrew Jackson into Florida. Jackson was on Spanish soil; he was only supposed to capture runaway slaves, but he did

Southern planter John C. Calhoun (below left), who was Monroe's secretary of war, started out as a strong believer in the union of states. He was a "war hawk" during the War of 1812, and an expansionist (he thought the country should grow as much as it could). In 1820 Calhoun agreed to the Missouri Compromise, even though it outlawed slavery north of latitude 36°30' (where is that?). But Calhoun changed. Maybe it was because he never got to be president (his big ambition); maybe it was because the price of cotton fell and with it some of the South's power. Calhoun began talking of states' rights and individual liberty (but not liberty for slaves). He became a Southern planter first and a U.S. citizen second. Keep John Calhoun in your mind. He will be very important in the national drama to come.

83

**Monroe was ambassador to France during the French Revolution. The Marquis de Lafayette and his wife had been imprisoned by the revolutionaries. Monroe's wife, Elizabeth (above), visited Madame Lafayette in prison on the day she was to be executed. The French, who wanted to stay friendly with America, were so impressed that they set Madame Lafayette free.**

**As the** power of the United States grew, the Monroe Doctrine became more and more important. But when Monroe first made his speech, a lot of people in Europe sneered at the idea of this upstart new country telling them what to do in South America. What really kept most European nations away was fear of the mighty British navy.

more than that. He had learned to fight as the Indians fought. He burned villages and destroyed crops. He captured, killed, and humiliated the Seminoles. His White Stick Creek Indian allies fought with him. More than half his soldiers were Creeks. It was the First Seminole War. Osceola, who was now 14, was taken prisoner (but later released).

Some Americans were upset. They thought Jackson had gone too far. The Spaniards were really upset. John Quincy Adams was President Monroe's secretary of state. He offered Spain a deal. He said the United States would pay Spain $5 million for Florida. (Spain could still hold Texas and California and other western regions.) The Spaniards signed John Quincy Adams's treaty. They had no choice; Spain was too weak to fight. General Andrew Jackson was named governor of the new U.S. territory: Florida. It was 1821.

Lots of people in the United States couldn't wait to move to Florida. But what about the Seminoles? They had to make way for the white settlers. They were forced to move south to an inland reservation on sandy, barren land where crops hardly grew at all. Soon many were starving. The young man, Osceola, went south with the others. He was now a military chief, a *tustenugee*, a kind of policeman.

Florida wasn't the only place where Spain lost out. Spain and Portugal could no longer control their colonies in South and Middle America. One by one they had revolutions. They became independent nations. As soon as that happened, other European countries began to look greedily at those new Latin-American nations. James Monroe and John Quincy Adams decided something needed to be done to keep Europe out of the Americas.

In December 1823 President Monroe gave a speech to Congress. He said that the American continents were closed to other nations. He told the European countries that they were not welcome to look for colonies in this hemisphere. The United States will not interfere in Europe's affairs, said Monroe, so Europe should keep its hands off America. That speech is very famous. What he said is called the *Monroe Doctrine*. It has been American policy since the days of James Monroe.

The years that Monroe was president have been called an "era of good feelings." Most things were going well in the country. The old fight between the Hamiltonian Federalists and the Jeffersonian Republicans seemed to have died down.

But it is hard to have people and politics without having arguments, and, before long, people were fussing about politics again. And now there were new parties and new arguments.

# 17 JQA *vs.* AJ

John Quincy Adams aged 16. He was already an experienced diplomat.

John Quincy Adams wanted to be president…and so did Andrew Jackson. Their election battle in 1824 was hot-tempered. When Adams won and became our sixth president, Jackson's supporters were very angry. They believed Jackson had been cheated out of the presidency. (He wasn't.) They weren't interested in extending the era of good feelings. Now there was an era of political grouchiness.

John Quincy Adams was a lot like his Puritan ancestors: honest, intelligent, virtuous, and hardworking. He was as learned as any man

In this 1824 cartoon the betting on the race for the White House is hot and heavy. Balding John Quincy Adams is just ahead of his opponents, William H. Crawford of Georgia (left), and skinny, uniformed Andrew Jackson (right).

## Dear Sir

*When he was nine, John Quincy Adams wrote this letter to his father:*
Dear Sir:
I love to receive letters very well, much better than I love to write them. I make but a poor figure at composition. My head is much too fickle. My thoughts are running after bird's eggs, play and trifles, till I get vexed with myself. Mamma has a troublesome task to keep me a studying.…I wish, sir, you would give me in writing some instructions with regard to the use of my time, and advise me how to proportion my studies and play, and I will keep them by me, and endeavor to follow them.

With the present determination of growing better, I am, dear sir, your son,
JOHN QUINCY ADAMS

**John Quincy Adams**
PRESIDENT, 1825–1829

JQA wrote: "My life has been…marked by great and signal successes which I neither aimed at nor anticipated."

who ever sat in the White House. It was said that he could write in English with one hand while he was translating Greek with the other. But he wasn't much fun to have around. He was just too serious.

John and Abigail Adams had made sure their brilliant son was well educated and trained to serve his country. When he was 14 years old he went to Russia as secretary to the U.S. ambassador. (Young Adams spoke fluent French, the language of the Russian court; the American ambassador could hardly sputter a word of that language.) After that, John Quincy was always studying or serving his country. He had some of the best ideas any president has ever had, but he didn't know how to deal with people. That brain of his scared some people. In addition, many Southerners feared that Adams wanted to end slavery—and they were right. So there were many who opposed him. But John Quincy Adams was a very capable president. It was shyness that kept him from relating well to people. He wasn't pompous or stuck up. He just didn't know how to chat.

Newspaper reporter Anne Royall found a way to get him to talk. She discovered President Adams swimming in the Potomac River and sat on his clothes, which were on the riverbank. There wasn't much the president could do but answer her questions.

J. Q. Adams may have been the best prepared president ever, but some found him tiresome. President John Adams hadn't been able to get himself reelected for a second term, and neither could his son.

## The Bride Wore Blush

John Quincy Adams was a difficult husband as well as a difficult president. His wife Louisa once complained that "hanging and marriage were strongly assimilated." On one occasion Louisa was given some rouge by the queen of Prussia, who thought she looked pale. When Louisa decided to wear the rouge to an opera performance, she found that it "relieved the dullness of my homemade dress and made me look quite beautiful." When JQA saw it, he got angry, grabbed a towel—and wiped it off. Louisa was very upset. A few months later she put the rouge on for another party and this time refused to take it off.

# A Virginia Visit

*In 1810, 21-year-old Elijah Fletcher got on a small horse and left his home in Ludlow, Vermont. He was heading south, to Virginia, to teach school. Elijah, the sixth of 15 children, wanted to keep his family informed of his adventures. Here are his words (and spelling) from letters he wrote home:*

When there was no convenient waterway, tobacco barrels were rolled to market.

I have seen more log houses since I left home than I ever saw in Vermont. I find it is the principal manner of building in these parts, especially in New Jersey and Maryland. In Pennsylvania there are many, and in fact most all, stone houses and barns....I found N.J. as little inhabited as the wilds of Siberia. Two miles from Baltimore the land is covered with shrub-oaks and uninhabited.... There is no country, I believe, where property is more unequally distributed than in Virginia. We can see here and there a stately palice or mansion house: while all around for many miles we behold no other than little smoaky huts and log cabins of poor, laborious tenants.

*After teaching for a year in Alexandria, Virginia, Elijah (who had sold his small mare) got in a stagecoach and headed west. Now he was going to teach at a school near Charlottesville.*

We passed through Orange County, within 4 miles of Montpelier, President Madison's seat: saw the manner of teamsters travelling by carrying their own and their horses' provision and at night kindling up a fire beside the road and making the open air the house of entertainment: saw also the manner of rolling tobacco by putting a pole for an axletree through the middle of the hogshead...putting in horses and so rolling it upon its hoops 2 or 3 hundred miles to market....

About 7 o'clock in the morning we hove in sight of the famous Monticello....We arrived...at Char., a village of about 400 houses, courthouse, and good taverns. [Elijah had a letter of introduction to Esquire Garland.] I gave him my letter: found him a plain, jovial, unceremonious planter, his wife friendly and agreeable, and his children, some large, some small, 10 or 11—dressed in farmer style—and appeared more like robust, hearty Green Mountain boys than any I had yet seen in Va.

Monday 6th, I rode back to Charlottesville with Esquire G. on the purpose of visiting Monticello...but the hard and constant rain frustrated our intention....Wednesday 8th I started again for Monticello. Mr. Kelly, when I got to Char., went with me. When we arrived at the foot of the hill, we wound a side way, circuitous course to avoid the steepness in getting to the house which was immediately upon the top of the mountain. We rode up to the front gate of the dooryard, a servant took our horses. Mr. Jefferson appeared at the door. I was introduced to him and shook hands with him very cordially. We went into the drawing room. Wines and liquers were soon handed us by the servant. He conversed with me very familiarly, he gratified my curiosity in showing me his Library, Museum of curiosities, Philosophical apparatus &c. Mr. Jefferson is tall, spare, straight in body, his face not handsome but savage.

Monticello—Italian for "little mountain"—was named for the hill it stood on.

# 18 A Day of Celebration and Tears

Benjamin Rush was a scientist and doctor who served in the Revolutionary War, and a signer of the Declaration of Independence. He was also a friend of Jefferson and Adams—and it was he who reconciled the two after so many years.

Before I go on I want to tell you of something that happened on the Fourth of July in 1826. It was while JQA was still president. Do a little arithmetic and you'll see that day was the 50th anniversary of the signing of the Declaration of Independence.

Remember when those firebrands—Patrick Henry, Sam Adams, and Tom Paine—got people to want to break away from England? Remember John Adams and Ben Franklin, who said it should be done with a declaration? Remember how Thomas Jefferson chose just the right words for that declaration? And how John Adams made sure that it was signed by the men of the Continental Congress? (Those brave men became traitors to England when they put their names on that declaration. The penalty for treason was death.)

Well, 50 years later, everyone remembered those great men and those great events. People looked at the young nation and were proud of what it had accomplished. Oh, there were problems, plenty of problems; but to most of its people, America seemed the most exciting place in the whole world.

The United States had been born of an idea, an idea that struck thinking people like a bolt of lightning. It was the idea found in those simple words from the Declaration: *all men are created equal*. No nation before had made that its creed.

Men, women, and children from all over the world were pouring into America. They wanted to be part of a nation where people were believed to be *endowed by their Creator with certain unalienable rights, that among these are life, liberty, and the pursuit of happiness.*

On July 4, 1826, the two men most responsible for the Declaration—Thomas Jefferson and John Adams—were still alive. They had been good friends when the Declaration was written. Later, they became political enemies: one a Federalist, the other a Republican. And then, after both retired, a friend of each visited John Adams in Massachusetts. "I love Thomas Jefferson and I always shall," said Adams. "That is enough for me," said Jefferson, when he heard of the remark. He sat down and wrote Adams a letter.

And so began a correspondence—a series of remarkable letters about their ideas, and memories, and hopes—letters that you can find in the library.

As that 50th Independence Day approached, Americans were especially proud of those Founding Fathers and of the great Declaration. They were pleased that John Adams's son was president. Celebrations were planned across the nation.

The mayor of Quincy, Massachusetts, wrote to John Adams and asked him to take part in a special ceremony. But Adams was 91, and not feeling well. He wrote these words to the people of Quincy:

> *My best wishes in the joys and festivities and the solemn services of that day, on which will be completed the fiftieth year from its birth, the independence of the United States: A memorable epoch in the annals of the human race, destined, in future history, to form the brightest or the blackest page according to the use or the abuse of these political institutions by which they shall, in time to come, be shaped by the human mind.*

Now wasn't that typical of old John Adams! Even at 91 he was still warning his fellow citizens that things could go wrong; that it was up to them to use their minds for the good of all.

In Washington, D.C., the citizens were also planning a celebration. Washington's mayor wrote to Thomas Jefferson and asked if he would participate. But Jefferson was 83 and ailing. He wrote back saying he wished he could come, but he was not well enough. Then he said:

This bust of Jefferson was made from a plaster life mask the year before his death. Jefferson was deeply in debt —his friends even held a lottery to help his finances.

**In a letter** to Thomas Jefferson written on December 8, 1818, John Adams wrote, "While you live, I seem to have a Bank at Monticello on which I can draw for a letter of friendship and entertainment when I please."

**What did** Thomas Jefferson mean when he said, "the mass of mankind has not been born with saddles on their back, nor a favored few booted and spurred, ready to ride them"?

**When** Jefferson wrote, "These are grounds of hope for others," the "others" he was talking about were slaves.

*All eyes are opened, or opening, to the rights of man. The general spread of the light of science has already laid open to every view the palpable truth, that the mass of mankind has not been born with saddles on their back, nor a favored few booted and spurred, ready to ride them....These are grounds of hope for others. For ourselves, let the annual return of this day forever refresh our recollections of these rights, and an undiminished devotion to them.*

And that was just like Jefferson, to still be writing inspiring words.

Finally, when the Fourth of July came, Thomas Jefferson lay in his bed at Monticello. He asked, "Is it the Fourth?" and soon breathed no more.

In another bed, on that same day, this one in Quincy, Massachusetts, John Adams whispered, "Thomas Jefferson survives," and then he was dead.

Messengers on horseback set out from Massachusetts carrying the sad news south. Messengers were already on their way north from Virginia with their sad news. They met in Philadelphia, the city where young Adams and young Jefferson had worked together on the great Declaration. There, in the shadow of Independence Hall, the couriers exchanged their messages.

The great artist Gilbert Stuart was old himself when he painted this portrait of 90-year-old John Adams. By then Adams was grumpier than ever—but the two old men thoroughly enjoyed the painting process and got along very well.

# 19 Old Hickory

**Jackson's coming!**
**CLEAR THE COURSE!!**

When the old Hero hove in sight of the Capitol at Washington,

**March 4, 1829,**

he waved his hand and ordered all bargainers and billiard players, to leave the Cabinet. Old *Ebony* and *Rush* started one way, and *Clay* and *Peter B. Porter* another, all on horseback, Jehu-like, as will be seen above. The old General locked up the treasury, and directed that John Binns' salary should be stopped. A terrible rout followed. The whole air was rent with shouts of

Clear the Course ! Clear the Course !
The old Hero's coming ! *Farewell, a long farewell,* to the " CORRUPT HOUSE OF BRAINTREE !

**After Jackson defeated JQA, this flier summed up the glee of his supporters, ready for a self-made man in the White House.**

There is something you might have noticed about the first six presidents: they were all from Virginia or Massachusetts. There is something else too. They were all aristocrats. They had been born into successful, prosperous families. That gave them the time and opportunity to be well educated. John Adams and his son John Quincy Adams were both Harvard graduates. James Madison went to the College of New Jersey (Princeton). Thomas Jefferson and James Monroe attended the College of William and Mary. Only George Washington was not a college man; still, he was well read…and rich.

Now how do you think you would feel if you were living in 19th-century America in Tennessee and you were poor? Do you think you would have a chance to be president?

After Andrew Jackson was elected the seventh president, you would know you had a chance. If Andy Jackson could be president, then any white male born in the United States could be president.

Jackson was born in 1767 in a log cabin on the border between North and South Carolina. His parents were poor Scotch-Irish farmers. The Scotch-Irish were people who had moved from Scotland to Northern Ireland to the United States. Andrew's dad died just before he was born; his mother had to move with her three sons into her sister's home. There were 11 children in the house, and you know how kids are. It was a noisy place, and Andy soon learned to speak out, and

| Andrew Jackson |
|---|
| PRESIDENT, 1829–1837 |

**Andrew Jackson** was not much of a speller, and he often spelled the same word different ways. As he said, "It's a poor mind that can think of only one way to spell a word."

**Andrew** Jackson's democracy was only for white males. It would be many years before blacks, Indians, Asians, and women were allowed to vote. However, Jackson's presidency paved the way for all peoples.

**Andrew Jackson** was shot in the arm during a battle. The army doctor wanted to amputate his arm. Jackson decided to get a second opinion. He went to a Cherokee medicine man, who saved his arm.

Jackson believed in a strong executive branch and was supposed to be an admirer of Napoleon. "I feel like Napoleon, I thirst for glory!" he is saying. "Down with the Senate!" Do you think he really said that?

**In 1833,** Andrew Jackson became the first American president to ride on a train. He traveled on the Baltimore and Ohio, the first public railroad in the United States.

wrestle, and make the older boys respect him. He also learned to read.

In those days, many farmers could not read. So most towns had public readers—people who read the newspapers aloud so everyone would know the news of the colonies.

When Andrew Jackson was a boy he was asked to be a public reader. He said he was "selected as often as any grown man."

He was nine, in 1776, when he read aloud a declaration that had been written in Philadelphia. It was signed by John Hancock, president of the Continental Congress. It said, *We hold these truths to be self-evident, that all men are created equal,* and it also said that *these United Colonies are, and of right ought to be, free and independent states.*

Can you imagine the excitement that everyone felt when they heard those words? Can you imagine reading them to a group of grownups?

Four years later, Andrew joined the South Carolina militia to help fight for that Declaration of Independence. He was captured and ordered to clean a British officer's boots. When Andrew wouldn't do it, the angry officer slashed him with a sword. Jackson carried the scars for the rest of his life.

Andrew Jackson's older brother was killed fighting in that Revolutionary War. Andrew and another brother were taken to a military prison. They both got smallpox; his second brother died. Then his mother died. At 14 Andrew Jackson was an orphan. "I felt utterly alone," he said.

He learned to take care of himself. He had to. He was a fighter with an explosive temper who liked to have a roaring good time. Sometimes he drank and gambled. But he learned from his mistakes. People were attracted to him. He was smart and honest and fun to be around. He tried being a schoolteacher, but that didn't work out. Then he studied law and became a lawyer. When he was 21 he was appointed attorney general for the region that would soon become Tennessee. By the time he was 30 he owned two large plantations near Nashville.

Andrew Jackson was a man of action, a born leader who was always doing things and going places and changing the world he lived in. He served in Congress; he was a judge, a general, and a military hero. "He came into national party politics like a cyclone from off the western prairies," wrote a professor named Woodrow Wilson (who would become president himself). Jackson formed a new political party: the Democratic Party.

But when Harvard University gave an honorary degree to Andrew Jackson, John Quincy Adams was so horrified—he was a Harvard graduate and thought his school was disgracing itself—that he refused to attend the ceremony. JQA called Jackson "a barbarian and savage who can scarcely spell his own name." That "barbarian and savage" became an astonishingly popular president.

He was tall and lean and stood straight as the barrel of a rifle. His eyes were blue, bright blue. His thick hair was the color of sand, although it turned silver long before he became president. His soldiers called him "Old Hickory," because they said he was strong as a hickory tree. In his portrait you can see a kind face. It was also a sad face. His wife, Rachel, whom he loved dearly, died just before his inauguration.

What an inauguration it was! The people—ordinary people—had elected one of their own to be president. They wanted to be there to see him take office. They wanted to celebrate with him.

And so they did. Some came from 500 miles away. It seemed to people in Washington as if everyone in the West had come to town for the big day. And they all wanted to get into the White House—at the same time. They poured in through the doors in their buckskin clothes and muddy boots, and they climbed on the satin chairs and broke glasses

**Jackson was** 6 feet 1 inch tall, and, when he was president, weighed 140 lbs. An observer in 1815 described him as having "limbs like a skeleton." His pretty wife Rachel was said, by the same observer, to be "a short fat dumpling." He smoked and chewed tobacco, and that may have caused the hacking cough and headaches that pained him.

Rachel Jackson dreaded going to the White House. She said, "I had rather be a doorkeeper in the house of God than live in that palace." She never did live in that "palace."

Scholarly John Quincy Adams's departure from the White House was the end of an era.

**Andrew Jackson** bought 20 spittoons for the East Room of the White House for $12.50 each. Some people thought that a great waste of government money; others said it would save the White House carpets.

**Voting in** the new republic wasn't by secret ballot. You spoke out your choice, or raised your hand, and everyone knew your vote.

and spilled orange punch and pushed and shoved each other. President Jackson had to go out a back door to get away from the mob. Finally someone thought to bring buckets of punch onto the lawn and that got the crowd out of the President's House.

Some people remembered President Washington's receptions, where men wore gloves and silver buckles and talked softly. Those were the good old days; how comfortable they seemed. This modern world of Andrew Jackson and his friends would be the end of the United States, some said. That old Federalist, Chief Justice John Marshall, swore in President Jackson. It was said that Marshall would just as soon have sworn in the devil.

Mobs would take over; life would be awful—so the aristocratic leaders thought. But it didn't happen. Andrew Jackson did change the presidency—it was never the same again. Most people think he made it stronger.

It helped that he had good manners, natural manners. People who thought they would be angry at him ended up being charmed.

For many of the earlier presidents, democracy had meant government *for* the people. For Jackson, democracy meant government *by* the people. "Let the people rule," was his motto. And, ever since Andrew Jackson's time, we have.

**"I accept the office given me by the free and unbiased suffrage, of a virtuous people, with the feelings of the highest gratitude," said the new president, Andrew Jackson.**

# 20 Yankee Ingenuity: Cotton and Muskets

**Sam Slater arrived in America with only the clothes he stood up in. When he died, he was worth $1.2 million.**

Back in colonial times, Americans raised most of the food they ate and made most of what they wore. They spun their own yarn, wove their own cloth, and stitched their own clothes. They dipped candles and built tables and chairs. Wealthy colonists who wanted fancy dishes, fine cloth, elegant furniture, or handsome books sent to England for them. Most manufactured goods were made in England; raw materials came from the colonies.

It was a system that worked well. America provided lumber, pitch, tobacco, cotton, and grains. England took those raw materials and turned them into usable products that could be sold around the world.

During the American Revolution the system stopped. Wham! Suddenly there was no place to send raw materials and no supply of fine goods. What did the colonists do? They used their heads. They looked for new markets for their raw materials. Their ships sailed to far-away places: to Spain, to China, to India, to Turkey.

After the war the new United States began trading with England again. But American society was changing. We were now a democracy with a strong and growing middle class. It wasn't only the very rich who wanted to buy things. Ordinary people wanted them, too.

In England something was happening that could make that possible. That something was an "industrial revolution." Let me explain. It is the end of the 18th century, and if you want a new shirt, this is what you have to do:

Take some wool, or flax, or cotton, and sit down at a spinning wheel. Try wool. You have to turn that sheep's wool into yarn. That takes

**Instead of** selling their farm products in the market, some Americans were now selling their time in a factory. (But, throughout the 19th century, most Americans remained farmers.)

## River Power

The Blackstone River was named for the Rev. William Blackstone, who came to Rhode Island in 1635—a year before Roger Williams arrived. A hundred and fifty-five years later, it was where the American industrial revolution began. The river twists and turns and drops and tumbles—from Worcester, Massachusetts, to Providence, Rhode Island. That river action is perfect for industry that depends on waterpower. Falling water can turn wheels that spin cotton, forge iron, and grind grain. The Indians had used the Blackstone for drinking water and fishing; industry made it among the most polluted of America's rivers.

Long, narrow fleeces of combed, raw cotton were fed into a "drawing machine" (top) to make them extra smooth and fine. After spinning, the yarn was wound onto the looms' giant frames for weaving (bottom).

carding (combing) and then spinning. It is a slow process. Since you have been working all winter, you do have a supply of yarn. How about dipping the yarn in indigo (blue) dye? Now, unless you plan to knit your shirt, you're still not ready to make it. You need to sit at a loom and weave the yarn into cloth. When you've done that, then, finally, you can get a scissors, cut out a pattern, and sit back with your needle and thread and start sewing. Do you now see why you have only two shirts—one for everyday and one for church?

Here is Elizabeth Fuller's diary for a few days in 1791.
*Aug 16. I picked blue wool.*
*Aug. 17. I broke blue wool.*
*Aug. 19. I carded blue Wool. Ma spun.*

And here is what Lucy Larcom had to say (you'll read more about Lucy in a minute):

> I think it must have been at home, while I was a small child, that I got the idea that the chief end of woman was to make clothing for mankind....I suppose I have to grow up and have a husband, and put all those little stitches into his coats and pantaloons. Oh, I never, never can do it!

Well, something was happening in England that was changing that made-at-home way of making things. It was a revolution—an industrial revolution (although no one called it that for a while). It was a new

Slater's Mill at Pawtucket (below) launched the Industrial Revolution in America. It changed forever the way textiles were made. For a while handworkers such as ladies' dressmakers (bottom) were still needed—but soon their jobs, too, would be adapted to factory production and done more cheaply.

system of organizing work, based on new ideas in science and technology and business.

Things once made at home were being made faster, and sometimes better, in factories. Tasks were divided in new ways. People began working in teams, and that was much more productive than working alone. It was machinery that made it all possible. Americans wanted some of those machines.

The English weren't about to share their new knowledge. They wanted to be the only ones with the machinery that made factories possible. They wanted to keep the Industrial Revolution in England. They wouldn't let anyone who worked in a cotton factory leave England.

97

Francis Cabot Lowell (inset) built his textile factory in Waltham, Massachusetts, a few miles west of Boston. His workers' houses had a homey, comfortable look. But most mill towns were dreary.

**When Sam** Slater came to Providence, he boarded with a Quaker family, the Wilkinsons. Hannah Wilkinson was the eldest daughter, and she and Sam fell in love and were married. Some years later, Hannah became famous in her own right: she developed fine cotton sewing thread—the kind you buy on spools—which she made from the fine cotton yarn that Sam manufactured.

Some Americans offered a big reward to anyone who could build a cotton-spinning machine in the United States. Samuel Slater, a young apprentice in a cotton factory in England, had a remarkable memory. He memorized the way the machines were built. Then he ran off to London. In London he pretended to be a farm worker. He didn't tell anyone he had worked in a cotton mill. It was 1790 when he sailed for America; he brought the key to the Industrial Revolution with him.

Slater built a small factory next to a waterfall on the Blackstone River at Pawtucket, Rhode Island. (Moses Brown and William Almy were his partners. They provided the money.) Waterpower turned the machines that spun cotton fibers into yarn. Soon there were spinning mills beside many New England streams. (Women working on handlooms in their homes wove Slater's yarn into cloth.)

Now that factories could turn cotton into yarn—quickly and easily—you can see there would be a great demand for raw cotton. Anyone who could grow cotton would make a lot of money. Cotton grew very well in the Southern states.

The cotton that grew in the coastal region was easy to use. It was called long-staple cotton and it had seeds that fell right off the cotton bolls. But the tidewater coastal lands were in poor shape. There wasn't much good land left. People didn't practice scientific farming. They often destroyed land by growing the same crops year after year. Then, when the land was no longer productive, they moved on.

Short-staple cotton was the only cotton that would grow inland. However, short-staple cotton has lots of dark seeds, and those seeds stick to the cotton bolls. You can't spin cotton that is full of black seeds. It took a worker all day to remove the seeds from just one pound of cotton. If only there were an easy way to get rid of those seeds…

Eli Whitney heard all about that problem when he came to Savannah, Georgia, to take a job as a teacher. Whitney, a New Englander with an inventive mind, had just graduated from Yale College. It took him very little time to come up with a simple machine that removed seeds from cotton. He called it a "cotton engine"—the name was soon shortened to *cotton gin*. Instead of taking all day to remove seeds from a pound of cotton, a worker with a cotton gin could clean 50 pounds of cotton in a day—and clean it better than he ever could by hand.

The invention of the cotton gin, in 1793, did something that no one expected: it encouraged slavery.

The South had been having economic problems. Slavery wasn't as useful as it had been in the early colonial days. Tobacco had used up the soil. There wasn't enough work for the slaves. Many slaves were set free because owners no longer wanted to feed and house them. Thomas Jefferson and the other Founders thought slavery might gradually disappear.

Eli Whitney's cotton gin changed things—really changed things. If you could grow a lot of cotton you could get rich. So Southerners looked for land to grow cotton and workers to plant and harvest it. Slaves became very valuable again. Whitney didn't mean it, but his invention helped turn the American South into a slave empire. It made the South into a land of cotton. It kept it rural.

At the same time, the North was becoming urban and industrial. It began in earnest after 1810, when a Boston businessman named Francis Cabot Lowell took a trip to England. While he was there he visited a clothmaking factory. (Remember, in the U.S. cloth had to be made on handlooms.) Lowell was able just to look at the English power looms and understand the way they were built. No one believed that could be done. When he came home to America he built a factory that was even better than those in England. Lowell's factory had machines for both spinning and weaving. He took cotton fibers and turned them into finished cloth—all in the same building. Even in England, no one had done that.

Once you get started with machines and technology, one invention seems to lead to another. In the old days most things were made from

Child workers could crawl under machines to oil and clean them. Children's small fingers were nimbler when broken threads had to be tied. And children didn't have to be paid adults' wages.

**In the** 18th century, only the very wealthy wore cotton clothes. Factories changed things. They could turn out products quickly and efficiently. They made goods affordable. Because of people like Eli Whitney, Sam Slater, and Francis Lowell, most Americans were soon wearing cotton.

99

The plantation on which Eli Whitney (top) designed the cotton gin belonged to the widow of a fellow New Englander: Gen. Nathanael Greene, who had been the Revolutionary army's quartermaster.

*Why do factory goods cost less than handmade ones?*

**According to** the law in Massachusetts, children were allowed to work only nine months a year, so that they could go to school for the other three months.

start to finish by one worker. A musket, for instance, was made by a gunsmith who would make guns one at a time. No two muskets were exactly alike. If a musket broke you had to find a gunsmith to repair it. You can see that might be a problem on a battlefield.

Eli Whitney began making muskets with interchangeable parts. In a factory one person could make all the stocks and another all the barrels. The parts from one musket would fit every other musket. It was very efficient. It had another big advantage. Can you see what it was?

Suppose a part broke. If parts are interchangeable you can fix a musket by taking parts from another musket. Think of the advantage of that on a battlefield. It was a simple idea—like the cotton gin—but it changed industry. Eli Whitney wasn't the first to design interchangeable parts. But when he showed his muskets to a group of politicians that included President John Adams and Vice President Thomas Jefferson, he made the idea popular.

Slater, Lowell, and Whitney helped bring the factory system to America. There were big advantages to the system, but disadvantages too. The skilled craftsman, who took pride in his work and used his mind as well as his hands, became rare. Unskilled workers could now do things that only artisans had been able to do in the past.

Factory goods cost much less than handcrafted goods. That meant that ordinary people could afford things they had never been able to buy before. That made life better for most people. But not for everyone.

Work in the factories was mind-dulling. Workers did the same task, over and over and over. "What can be expected of a man who has spent twenty years of his life in making heads for pins?" asked a Frenchman who visited some factories.

The air in the cotton mills was full of tiny, almost invisible cotton fibers that got into your lungs (and sometimes led to cancer or other ailments). Those new spinning machines and looms were big and powerful and had no safety devices. If a worker's hand slipped—she might lose it. The noise was deafening—some workers actually went deaf. Factory lighting was usually poor—that didn't help your eyesight.

Some of the workers in the factories were children. Some were as young as seven years of age. How would you like to work in a factory instead of going to school? Sounds good to you? Children often worked 10 or more hours a day.

Lucy Larcom was a real girl who lived in the 19th century. When she was little, a neighbor called Aunt Hannah, who kept a small school, taught her to read. They sat together near the kitchen fireplace. As Aunt Hannah twisted the thread on her spinning wheel, she pointed out words in the spelling book with a pin. Lucy was quick to learn. Soon Aunt Hannah taught her to read the stories in the Bible. Then Lucy went to school. Lucy loved school; she wanted to go on, to learn more, but...*alas, I could not go. The little money I could earn—one dollar a week...was needed by the family.* Lucy went to work in a factory.

Francis Lowell hired young farm women for his factory. Lowell housed them in dormitories and saw that they lived well and got fair salaries. But other factory owners took advantage of workers, especially women and children. They paid them poorly and made them work long hours.

Some millowners built mill villages. They provided whole families with jobs, houses, schools, churches, and stores. It sounded good—and sometimes it was—but it gave the owners control of the workers' lives. Then they could do almost anything they wanted. Millowners in Pawtucket lengthened the workday *and cut wages* at the same time. (The women weavers led a "turnout." It was one of the nation's first strikes.)

Those factory workers were taking part in two revolutions. The first was that industrial revolution; Slater and Lowell had helped bring it from England to America. The second, which was related, was a market revolution. That means the United States was going from a self-sufficient *farm economy* (where most families took care of their own needs, and rarely used money), to a capitalist *market economy*, based on jobs and money (where people earned wages and bought goods in markets and stores). These revolutions were just getting underway when Andrew Jackson became president. Once they got going they moved quickly and broadly. Revolutions do that—they change more than anyone ever expects.

> **On a** bone-chilling day in December 1790, Samuel Slater hooked up his spinning machine to the small water wheel in the Pawtucket mill. But the river was frozen over. Sam had to crawl onto the ice and break up the sheets around the wheel. The current flowed, the wheel turned—and the machinery began to hum. All that first winter, Sam had to keep breaking the ice around the wheel. It left him with bad rheumatism for the rest of his life.

This manufacturer was named for the Merrimack River, which ran through Lowell and powered all the machines.

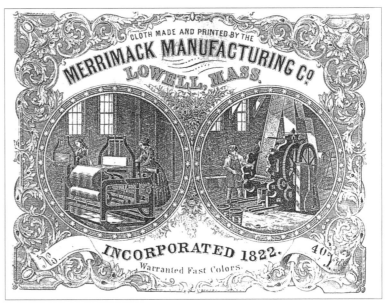

# 21 Modern Transportation

***Ingenuity*** (in-juh-NEW-uh-tee) means inventiveness—finding a way to get things done.

**By 1800 the mail stage could do the 200 miles from Boston to New York in two days—in good weather. In winter it might take a week.**

One thing leads to another: if you start making cloth, thousands of yards of it, you can't keep it all in New England. You have to send it to other markets.

If your ships go to China and bring back fine goods, you can't keep all those goods in Salem or Boston. If you grow grain in New Jersey, or forge iron in Pennsylvania, or make guns in Connecticut, you need to find ways to get your products to people who want to buy them. If you live in the West, you want to send your grain, furs, and cattle to eastern markets. How do you get your goods to market? How can you get cloth from Boston to Buffalo?

In the first half of the 19th century, roads were no answer. Picture this: ruts, holes, mud, stones, and when you come to a river—no bridge. Now you have an idea of the roads.

What was needed was modern transportation. Americans—who

**Bumpy corduroy roads such as this were an improvement on dirt tracks, which were soggy swamps in spring and choked with dust in summer.**

were becoming known all over the world for their ingenuity—soon came up with some answers. They were: canals, steamboats, railroads, and improved roads.

Let's start with the new roads. Actually they were pretty terrible—but in the 19th century they seemed exciting, and much better than the existing roads, which were usually just dirt paths. Do you know what *corduroy* is? Well, it is a cotton cloth with ridges and valleys. Some roads were made of round logs placed next to each other. They were called corduroy roads. Can you see why? What would it be like to ride on a corduroy road? Plank roads, made by placing flat wooden planks next to each other, were better, but not great. They quickly rotted away.

Road building was a new science. No one knew how to build good roads, so they experimented. Often the new roads wore out or washed away almost as soon as they were built. Still, some very useful roads did get finished. The best were macadam roads, built with a new process developed in Scotland (by a man named McAdam) using crushed stones and clay as a base and asphalt or tar on top.

**In 1807,** Fortescue Cuming wrote, "The travelling on these roads [in western Pennsylvania] in every direction is truly astonishing, even in this inclement [bad-weather] season, but in the spring and fall, I am informed that it is beyond all conception." (What he is saying is that an astonishing number of people were traveling.)

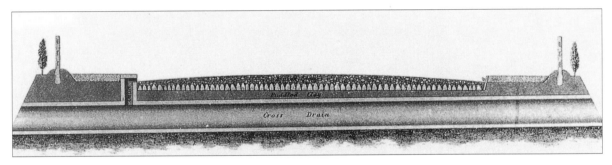

About 1806, some people with big ideas decided that we needed a road that would go across the country—well, at least from the East Coast to the Mississippi, which seemed across the country to most Easterners then. (Remember, the Louisiana Purchase was made in 1803. Hardly anyone knew what was beyond St. Louis.) That very long road was called the National Road and was to be paid for by the federal government. Does that sound like a good idea? Well, it didn't seem that way to everyone. It caused a whole lot of controversy.

The people in the West wanted it—they really wanted it. But many Easterners said, "Why should our tax money go for a road out to that wilderness?" In the South, people were shouting about states' rights. They didn't think the national government should pay for roads. If that happened, even states that the road didn't go anywhere near would have to help pay for them. President James Monroe said that it was

The Romans and Aztecs knew how to make good roads, but their skill was forgotten until the end of the 18th century. Then engineers began to build stone roads with drainage and a slope, or *camber*, for water runoff.

**Some people** called the National Road the Cumberland Road.

Ahead of the settlers and pioneers went the surveyors, mapping routes and proposing roadways. Indians called the surveyors "land stealers."

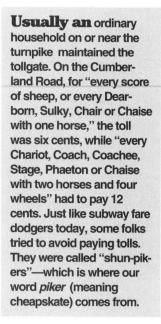

unconstitutional. But finally the National Road was begun. By 1818 it stretched from Cumberland, Maryland, to Wheeling in western Virginia. Then powerful Senator Henry Clay got involved. He wanted to see the road extended, and it was. By 1833 it went to Columbus, Ohio; by 1850 it was at Vandalia in central Illinois.

Over its wilderness route passengers squeezed shoulder to shoulder in hard-backed, leather-seated stagecoaches. That was a lot better than walking, which is what many did alongside their packed wagons. The road carried all kinds of traffic: mule-drawn carts heaped with farm produce; big, horse-drawn vans stuffed with bales of southern cotton going to northern mills; northern factory products heading south; and wagonloads of immigrants, still speaking foreign tongues, bound west to destinations they could not even imagine.

Before the National Road was built it took four weeks to travel from Baltimore to St. Louis. On the road, if you traveled without stopping, you could make it in four *days*. One traveler (his name was Charles Fenno Hoffman), who went the whole way on horseback, wrote this:

> *It appears to have been originally constructed of large round stones, thrown without much arrangement on the surface of the soil, after the road was first levelled. These are now being ploughed up, and a thin layer of broken stones is in many places spread over the renovated surface....It yields like snow-drift to the heavy wheels which traverse it....There is one feature, however, in this national work which is truly fine, I allude to the massive stone bridges which form a part of it.*

New roads made it easier for Americans to travel, and to buy and sell goods. Some of the roads were built by private companies. The company would put sharp sticks—called pikes—on a movable rod that blocked the entrance to the road. To get on the road you had to pay a toll; the gatekeeper then turned the pikes.

Roads were expensive to build and maintain—there had to be a better and cheaper way to move goods and people.

Some people thought canals were the answer. Ben Franklin, back in 1772, wrote, "canals are quiet and very manageable." George Washington believed that canals were the wave of the future. He invested in the Potomac Canal system (near Washington, D.C.) and in the Kanawha Canal. Investors thought that canal would be like a major highway, taking goods and people from Virginia's James River across the Appalachian Mountains to the Ohio River. But the Kanawha Canal was never completed.

In New York, DeWitt Clinton decided a canal could be built from Albany to Buffalo, which meant from the Hudson River to Lake Erie. It would be named the Erie Canal. Look at a map. That was to be some canal!

A canal, by the way, is a big ditch. That's what some people called this one: "Clinton's Ditch." Many people thought it a crazy project. It

**The Erie Canal was built with only the most primitive machines, such as the horse-powered crane below. When Governor Clinton reached New York City to open the canal, he poured a keg of Lake Erie water into the bay.**

**The first** bicycles were seen at the beginning of the 19th century. In 1813 a celeripede (suh-LER-ih-peed) appeared in America. It had two wheels but had to be pushed with your feet, like a kiddie car. True bikes came in the 1830s, but with the pedals on the rear wheel and the seat over that wheel.

**The early** 1800s were years of fast growth. New words and expressions appeared in the language at that time—they were hurry-up expressions:
*like greased lightning*
*quick as a wink*
*in a jiffy*
*like a house afire*
*shake a leg*
*lickety-split*

would cost a lot of money and would be very difficult to build. The Erie Canal would have to traverse 360 miles, most of it through the wilds of New York State. There were steep hills to climb. Boats would have to get over those hills. To raise the boats, locks would be needed. A lock is like an elevator for water and boats.

When Thomas Jefferson heard of the project he said to a canal booster, "Why, sir, you talk of making a canal of 350 miles through the wilderness—it is little short of madness to talk of it at this day!" No question about it, it was a very ambitious project. But that didn't stop DeWitt Clinton or the many Americans who wanted to build the canal. It was 1817 when they set to work.

Thousands of laborers were needed. Ireland was having economic problems—people in Ireland were hungry. Canal workers got 50 cents a day and all the meat they could eat. That sounded good to many Irishmen, especially to those who'd just arrived in this country.

It was planned that horses or mules would pull most of the boats on the Erie Canal. No, the animals didn't have to swim. Workers built a towpath next to the canal. Boats were attached to ropes and towed by the horses.

Somehow it all got built. It was four feet deep and 40 feet wide. It was a manmade river, it was an engineering marvel! You could ride on a barge from the Atlantic Ocean, up the Hudson River, across the Erie Canal, and on to the Great Lakes. On the canal there were 83 locks to raise and lower you, your boat, and the water you were floating in.

The canal was officially opened with a bang—actually, a whole chain of bangs. First, a cannon was fired in Buffalo and when that was heard down the canal, another was fired, and then another—in Rochester, Syracuse, Rome, and Utica—until they heard the blast in Albany, where they just kept the cannons going, on down the Hudson River. It took an hour and 20 minutes before the last cannon blasted in New York. Then, just for fun, they shot the cannons again, all the way back to Buffalo.

They didn't call it the "Big Ditch" anymore; now it was called the "Grand Canal" and sometimes "Clinton's Wonder." DeWitt Clinton, who was now Governor Clinton, took the first ride from Buffalo to New York City. It had taken eight years to dig the Erie Canal. Clinton's trip took nine days. Clinton's canal boat was called the *Seneca Chief* and carried a portrait of the governor in a Roman toga. The boat behind was called *Noah's Ark*. It carried two Indian boys, two deer, two eagles, and a lot of other twos. Another boat held four raccoons, two wolves, a fawn, and a fox. When Clinton got to New York City he dumped a barrel of Lake Erie water into the Atlantic Ocean. Then there was a grand parade; church bells rang and people cheered.

Everyone could ride the canal. There were fancy passenger boats that served fine meals on linen tablecloths. There were flatboats with people, cargo, and animals jammed together. There were ordinary rafts. You could go on a slow boat, at two miles an hour, and pay a penny and a half a mile. Or you could whiz along at four miles an hour and pay five cents a mile. Towns grew up around the canal; it made life better for people. Before the canal was built, it cost $100 to ship a ton of grain from Buffalo to New York. By 1855 it only cost $8 on the Erie Canal. People packed their belongings and took the Erie Canal west; they moved to places like Indiana, Michigan, and Wisconsin. They went east, too; the canal helped make New York the country's largest city. Before long there was a canal frenzy throughout the nation. But no other canal was as successful, or as long, as the Erie Canal.

Canal laborers sang as they worked. Soon everyone was singing a song called "The Erie Canal."

The Erie Canal had 18 aqueducts—bridges that carry water over obstacles instead of people over water. Before railroads came, the canal was so popular that boats waiting for locks to open were often stuck in traffic jams.

*I've got a mule and her name is Sal,*
*Fifteen miles on the Erie Canal.*
*She's a good old worker and a good old pal,*
*Fifteen miles on the Erie Canal.*
*We've hauled some barges in our day,*
*Filled with lumber, coal, and hay,*
*And we know every inch of the way*
*From Albany to Buffalo.*

CHORUS
*Low bridge! Everybody down.*
*Low bridge! We're a-coming to a town.*
*You'll always know your neighbor,*
*You'll always know your pal*
*If you've ever navigated on the Erie Canal.*

*We'd better get on our way, old pal,*
*Fifteen miles on the Erie Canal.*
*You can bet your life I'd never part with Sal,*
*Fifteen miles on the Erie Canal.*
*Get us there, Sal, here comes a lock;*
*We'll make Rome 'fore six o'clock.*
*One more trip and back we'll go,*
*Right back home to Buffalo.*

*(Repeat CHORUS)*

# Old Heroes, Old Friends

It was 1824, 40 years after America fought its War of Independence, and a great French hero of that war, the Marquis de Lafayette (mar-KEE-duh-laff-fye-ET), was back in the United States for a visit. President James Monroe and the Congress invited him as "the Nation's guest." Lafayette was now an old man. Still, every American knew his name. So everywhere there were adoring crowds. A contemporary wrote that people were wearing "LaFayette boots—LaFayette hats…and LaFayette everything."

When he glided into Rochester, New York (on the new Erie Canal), bands played, cannons roared, and people stood and cheered. In Washington, D.C., Lafayette rode to the Capitol in a chariot with 24 young women (representing the 24 states). Each wore a white dress, a blue scarf, and a wreath of red roses. It was the grandest parade in the city's history. George Washington's tent had been set up in the rotunda of the Capitol (under the dome), just so Lafayette could walk in it once again.

At Yorktown, Virginia (the place where he had led troops to victory), people cheered and pushed. Lafayette waved from his carriage and then, suddenly, he saw someone he knew. The man he recognized was 76 years old—Lafayette had last seen him when he was 33. Still, according to a Richmond newspaper, the marquis "called to him by name, and [took him] into his embraces." The man he hugged was James Armistead Lafayette.

During the Revolutionary War, the Marquis de Lafayette had needed to know everything he could about the British army. Lafayette needed a spy in the British camp.

William Armistead, who was supplying food to Lafayette's troops, suggested someone who was intelligent, trustworthy, and believed in the American cause. His name was James, and he was Armistead's slave.

Marquis de Lafayette

*James Armistead Lafayette*

James understood the risks—spies are not treated with mercy. He volunteered to go into the British camp and see what he could learn. He began selling food to the British. The redcoats didn't suspect that he was carrying letters "from the Marquis into the enemy's lines, of the most secret and important kind."

In fact, the British were so impressed with James that they asked him if he would spy for them. Of course he said yes. It was a good cover and got him into the headquarters of the British general, Lord Cornwallis.

One day Lafayette gave James some torn pieces of paper. He was to take them to the British and say he had found them in the American camp. When pieced together, the papers said that General Daniel Morgan had brought his army into Lafayette's camp—which was just what Lafayette wanted Cornwallis to believe. (The American forces at Yorktown weren't ready to fight without Morgan—who had *not* arrived.) The ruse worked.

Soon after the war was over, Lafayette gave James a paper that said, "This is to certify that the Bearer by the name of James has done Essential Services to me while I had the Honour to Command in this State. His intelligences from the Enemy's Camp were industriously collected and most faithfully deliver'd."

Two years later the Virginia legislature declared James a free man. He took the name James Armistead Lafayette, moved with his wife and children to a farm, prospered, and bought three slaves. His hero, the marquis, said that if he'd known America would become a nation that supported slavery, he might not have fought in the Revolution at all. What do you make of all that?

# 22 Teakettle Power

**Besides the *Clermont*, Robert Fulton designed 16 other steamboats, a ferry, and a torpedo-firing submarine.**

Think about blowing a whistle. It takes lung power to do it. With that same power you can blow up a balloon or push a toy sailboat in the bathtub. Early in the 18th century it seemed to occur to several people that boiling water—steam—produced the same effect, and might be used as a source of power.

Have you heard a teakettle blow a whistle? Have you ever seen a teakettle blow off its spout? Steam is powerful stuff. Picture a huge boiler filled with water that is boiling furiously. The steam that it makes can move a boat or a train.

Some people—those with curious minds—began to think of ways to do that. Some Englishmen started the process, but quite a few Americans worked on the problem. William Henry, John Stevens, and John Fitch were American inventors—each built a steamboat—but each had some bad luck and the country wasn't quite ready for their ideas.

It was ready when Robert Fulton came along. Fulton was an artist, a good one, who had studied painting in London. Fulton was also an inventor. Some people called him an American Leonardo da Vinci. (Leonardo was one of the world's greatest artists: he was also an inventor.) People were exaggerating—Fulton's paintings weren't as good as Leonardo's—but his inventions were more important.

In 1807, Fulton's steamboat, the *Clermont*, steamed up the Hudson River the 150 miles from New York to Albany. It made that voyage in 32 hours, and that seemed astonishingly fast. (How fast? Do some arithmetic and find out how many miles per hour it went.)

**When I hear the iron horse make the hills echo with his snort like thunder, shaking the earth with his feet, and breathing fire and smoke from his nostrils (what kind of winged horse or fiery dragon they will put into the new Mythology I don't know), it seems as if the earth had got a race now worthy to inhabit it.**

—HENRY DAVID THOREAU, *WALDEN*

Fulton's *Paragon* (above, on the Hudson in 1808) was so grand—it had a paneled dining room and real bedrooms—that passengers unused to traveling in such style had to be reminded not to sleep in their shoes.

You have to understand that before the steamboat, boats floated down a river on the river's current. Can you figure out how they could go upriver against the current?

There was no easy way. Most traffic on a river was one way. If you really wanted to go in the other direction, you could use poles and push yourself (hard), sail (if there was enough wind), or be pulled by horses on a towline (many problems). On the Mississippi River, boats could sometimes make a mile an hour going upriver.

Fulton's boats were soon chugging up the Mississippi at 10 miles an hour (easily). By 1820 there were 60 steamboats on the Mississippi; by 1860, there were about a thousand.

Steamboats were efficient, fast, fun, and—dangerous! If steam gets trapped in a boiler and for some reason can't get out, the boiler will explode. That happened to a number of steamboats. They blew their lids and killed people when they did.

The same thing happened when steampower took to the rails. Now that was a good idea! Take a steam engine and have it pull a set of wagons rolling on a track. Hold onto your hat:

**Inventor Oliver** Evans designed and built a 15½-ton steam-powered vehicle and, in 1804, drove it through the streets of Philadelphia and into the river.

**Boiler explosions were common; safety devices were often inefficient, sometimes nonexistent. In 1871 New York's Staten Island ferry blew up, killing 60 people.**

we're going to whiz at 20 miles per hour. As steam engines improve, trains will go faster and faster.

But talk about dangerous! Besides exploding, the engines jump their tracks, and trains crash into each other. The trains are called "iron horses" and "teakettles on tracks." The engines are made of iron and have tall smokestacks and fireboxes. Passenger cars look like stagecoaches, and, like fancy coaches, are painted with bright designs. Because the cars are wide open, the colorful paint is soon covered with soot, and so are the passengers. And soot isn't the only thing the riders have to worry about: the engines give off sparks that fall on their clothes and get into their hair. On the first train in New York, passengers are kept busy putting out fires in each other's clothing.

But soon the design is improved, and the passenger car becomes a long, enclosed room with an aisle down the center and seats on either side. The first steam engines were fired with wood, but coal was found to produce a hotter flame. Either way, a fireman is kept busy lifting wood or shoveling coal to keep the fire blazing and the water boiling.

When the Baltimore & Ohio Railroad opens 13 miles of railroad track in 1830, horses are used to pull carriages on wheels. On a track, one horse can pull as heavy a load as 10 horses off the track. But the railroad's directors are not satisfied. They are forward thinkers. They want to try a new steam engine. British engineers who come and look at the track say sorry, the Baltimore & Ohio track has too many

curves for a steam engine. No engine can stay on that track, they say. Peter Cooper thinks differently. He is a Baltimore inventor and he owns an iron foundry. Later, he wrote, "I had naturally a knack at contriving, and I told the directors that I believed I could knock together a locomotive that would get around that curve."

Which is just what he did. He found a small engine, took the barrels from two muskets (for

**Peter Cooper (above) was a New Yorker who began with a grocery store and a glue factory. He built many machines besides gaily painted *Tom Thumb* (right).**

## Ride a Fire-Eating Monster

PAST

FIRST LOCOMOTIVE    THE ROCKET

PRESENT

George Stephenson,

*George Stephenson, the first to transport people by locomotive, and his* Rocket.

The first real locomotive—called the *Rocket*—was built in England in 1828 by George Stephenson. There had been earlier contraptions—in 1813, William Hedley's *Puffing Billy* hauled 10 wagons of coal at five miles an hour—but Stephenson's *Rocket* was the first real "iron horse." And *iron horse* was the name everyone gave to the powerful locomotives with big boilers for producing steam. Americans were intrigued with reports from England of these strong, new, snorting vehicles. Engineer Horatio Allen was sent to England to investigate. As soon as he saw them, he ordered four iron horses. When the first of them arrived in New York City, in May 1829, someone said it looked like a giant grasshopper.

That grasshopper/iron horse actually had a fierce lion's head decorating its front and was called the *Stourbridge Lion* (because it was made in Stourbridge, England). The creature was sent to Honesdale, Pennsylvania, for a trial run. Tracks there went over a creek on a wooden trestle bridge

that was 30 feet high. (Coal wagons had been pulled on the tracks.) No one wanted to ride an enormous engine over that bridge.

"The impression was very general," wrote Horatio Allen later, "that the iron monster would either break down the road or that it would leave the track at the curve and plunge into the creek." Allen decided that he would ride the engine—alone.

*As I placed my hand on the throttle-valve handle I was undecided whether I would move slowly or with a fair degree of speed; but believing that the road would prove safe, and preferring, if we did go down, to go down handsomely and without any evidence of timidity, I started with considerable velocity [speed], passed the curve over the creek safely, and was soon out of hearing of the cheers of the large assemblage present. At the end of two or three miles, I reversed the valves and returned without accident to the place of starting, having thus made the first railroad trip by locomotive on the Western Hemisphere.* (He meant the first locomotive trip in the Americas. It was a big achievement.)

*It looked like a grasshopper, but the* Stourbridge Lion *proved its might.*

pipes) and some other odds and ends, and built a little steam locomotive he named *Tom Thumb*, after the children's fairy tale. Then he invited the directors for a ride. "We started—six on the engine, and thirty-six on the car. It was a great occasion....We...made the passage to Ellicott's Mills in an hour and twelve minutes."

*Tom Thumb* worried a Baltimore stagecoach company. They didn't want competition from iron engines. So they challenged the owners of the train to a race. Their best horse-pulled railcar against *Tom*. Well, the horse was soon in the lead; then the engine built up power and pulled ahead. It was way ahead when an engine belt slipped out of place. The steam pressure fell and the train came to a stop. The people in the horse-drawn train laughed as they galloped by. They wouldn't laugh for long. Trains were the future.

Canals had a problem: they froze in winter. Stagecoaches and horse-pulled trains had problems too: they were small, and couldn't carry heavy freight. Besides, horses get tired and need replacing. Trains could be used year round. They could carry very heavy loads. By 1840, more than 3,000 miles of track had been laid.

Charles Dickens came to America in 1842 and rode on a train. Dickens was a famous English novelist and the author of the story of Scrooge and Tiny Tim called *A Christmas Carol*. This is part of what Dickens wrote about an American train ride.

> *On it whirls headlong, dives through the woods again, emerges in the light, clatters over frail arches, rumbles upon the heavy ground, shoots beneath a wooden bridge...suddenly awakens all the slumbering echoes in the main street of a large town, and dashes on haphazard, pellmell, neck-or-nothing, down the middle of the road. There—with mechanics working at their trades, and people leaning from their doors and windows, and boys flying kites and playing marbles, and men smoking, and women talking, and children crawling, and pigs burrowing, and unaccustomed horses plunging and rearing, close to the very rails—there—on, on, on— tears the mad dragon of an engine with its train of cars; scattering in all directions a shower of burning sparks from its wood fire; screeching, hissing, yelling, panting; until at last the thirsty monster stops beneath a covered way to drink, the people cluster round, and you have time to breathe again.*

By 1850 there were almost 9,000 miles of track in America; by 1860—the year before the Civil War—there were 30,000 miles of track. Traveling by train, at an unbelievable 30 miles an hour, you could go from New York to Chicago in only two days.

Cartoons like this one poked fun at teakettle power, but the artist didn't realize what was ahead. Steam power was changing the world.

## Before Steam Power There Was Horsepower

*If you were in school in the 1830s, you would probably be reading Peter Parley's First Book of History. This is what it said in its chapter on Maryland:*

The most curious thing at Baltimore is the railroad. I must tell you that there is a great trade between Baltimore and the states west of the Allegheny Mountains. The western people buy a great many goods at Baltimore, and send in return a great deal of western produce ....Now, in order to carry on all this business more easily, the people are building what is called a rail road. This consists of iron bars laid along the ground, and made fast, so that carriages with small wheels may run along them with facility....A part of the rail road is already done, and if you choose to take a ride upon it, you can do so. You will mount a car, something like a stage, and then you will be drawn along by two horses, at the rate of twelve miles an hour.

# 23 Making Words

**Sequoyah used some letters from our Roman alphabet for his system—others look different from ours.**

The Indian whose name was Sequoyah had talent. When he was small, he was able to draw amazing pictures of birds and horses and people. When he grew to manhood he became a silversmith and made fine jewelry: earrings, bracelets, and necklaces for the Indian men and women, and sturdy silver pieces to hang around the necks of their horses.

We don't know if he was happy as a silversmith; we do know that—once he saw something the white men could do—he became a different man. What he saw seemed too astounding to be believed.

A white man would take words, turn them into shapes, and scratch them on a slate or draw them on a piece of paper. Then another man could look at those shapes and say the first man's words. Was it magic?

Sequoyah could draw pictures of things he could see. But you can't see words. You can't touch words. How can they be pictured? Sequoyah was determined to find out.

Sequoyah was a Cherokee, a member of a proud nation of hunters, warriors, and farmers who seemed to do everything well. But no Cherokee could write his own language, because reading and writing were unknown to the Indian tribes. Indian stories and speeches were remembered and retold by orators, or storytellers, or singers. Sequoyah knew that often in retelling, words are changed. He thought the words of some of the great Indian leaders needed to be remembered just as they were first said. Would the white man's idea work with Indian words?

Sequoyah, being a Cherokee, had confidence. He believed he could find a way to write his people's words. But how was it to be done?

Suppose you didn't know how to read or write English or any

## A Cherokee Fable

The Cherokees tell the story of the tortoise and the hare. Their story may be different from the one you know. The tortoise got all her relatives to help out on the day of the big race. She stationed them along the way. Each time the hare turned a bend in the path—there was a tortoise. The hare thought it was the same tortoise. Finally he just gave up. He wasn't outraced, he was outsmarted.

other language; how would you begin to create a written language? Sequoyah thought up a logical system. He made marks stand for sentences. Soon he had so many marks and so many sentences he was lost. He knew he could never remember them all. Then he tried making marks for words. Then there were too many word marks to remember. Finally he made marks for sounds—and he knew he had a good system. He came up with 86 Cherokee sounds and an alphabet of 86 symbols. (No symbol had more than one sound, which made it easier to learn than languages like English where the *oo* in "good" and the *oo* in "food" are mighty confusing.)

Now figuring all this out took him a long time. His wife, Sarah, was angry because he wasn't making his beautiful silver jewelry. His family was becoming poor. His friends told him he was being foolish; some thought the symbols were dangerous witchcraft. They tried to get Sequoyah to stop. This is what Sequoyah said:

> *If our people think I am making a fool of myself, you may tell our people that what I am doing will not make fools of them. They did not cause me to begin and they shall not cause me to give up.*

Sequoyah was a determined man who knew he had something important to accomplish. If his friends couldn't understand that, well, he was sorry, but he wouldn't give up. But some Cherokees were so

**Sequoyah said** about finding a way to write down the Cherokee language: "I thought that would be like catching a wild animal and taming it."

Sequoyah and the *Cherokee Phoenix*, the first bilingual newspaper in any Indian language. Almost all Cherokees became literate in a few years. Even those living far from cities stayed informed of events in the larger world. For most Cherokees, unfortunately, the news was not good.

This painting of an Assiniboin chief before and after a visit to Washington, D.C., portrays some Indians' worst fears about contact with white people. Coming home, Wi-jun-jon is wearing unsuitable clothes, smoking a cigarette, drinking whiskey. Soon after his return Wi-jun-jon was killed by his own tribe for being a ridiculous braggart.

**The giant** redwood trees, called sequoias, and Sequoia National Park, in California, are named for Sequoyah, the brilliant Cherokee who devised a written language.

frightened they destroyed all his work. He had to start again. It took him 12 years to develop his written Cherokee language.

When his system was finished, in 1821, he taught it to his daughter. She was six. Then he invited his friends to visit. He sent his daughter across a field, beyond shouting distance. Then he asked a visitor to say some words—any words—and he made marks on a slate. The visitor took the slate to the child: she said the words!

The Indians tried it over and over again. Each time the child said the exact words.

Could it be trickery? Or witchcraft? Some Cherokees were still fearful. Sequoyah taught his nephew and other children. Try as they might, the visitors could never fool the children.

Now people were convinced. His wife was proud of him. The chiefs wanted to write the speeches of the great leaders—and Sequoyah did that. Then they learned they could "speak at a distance," by writing letters. They called the letters "talking leaves." Soon the Cherokees were printing a newspaper.

The problem with the Indians, said many white men and women, was that they were "savages" and "uncivilized." By that they meant that the Indians did not do and think as the white people did. But the Cherokees confounded the whites. Many of them did live as the whites did. Some Cherokees married white people. Some combined the two ways of life. They cleared the land, built big farms, planted apple and peach orchards, raised cattle and hogs, and lived in beautiful plantation homes. Some lived in European-style houses. Many owned slaves. Missionaries moved into Cherokee territory and converted some of the Indians to Christianity. They built schools. The Cherokees formed a government, wrote a constitution, and built a capital city with broad avenues and solid buildings. They became very prosperous. But the wealthier the Cherokees became, the more anxious other people were to have their land.

That Cherokee land stretched across the mist-covered southern Appalachian Mountains in a semicircle that reached from Kentucky to Alabama. Mostly the Indians built their farms and villages in the mountain foothills, where the land was fertile and the creeks brought abundant water. The immigrants who were pouring into the country from Scotland and Ireland and Germany wanted that same fertile land. After all, that was why they had come to the New World. They had been told there was land enough for everyone. But the land in the East was all taken. They would have to go to the frontier if they wanted land.

That was when they discovered a problem: most of that frontier belonged to the Indians. "Why should the Indians have so much?" they asked. "Isn't there enough land for everyone?"

So they pushed west. Some of them put up their cabins on Indian land. Many wanted to live peacefully with the Indians. Some didn't. In Europe they had read that the Indians were ruthless savages. Many believed that. A few hated Indians just because they were different. Some attacked and killed Indians. The Cherokees had a warrior tradition. They didn't know which white people were friendly and which ones weren't. They just knew their land and lives were being threatened, so they went on raids and killed white people and burned their cabins and farms. The settlers didn't know which Indians could be trusted and which ones were killers, so they went on raids and killed Indians and burned their homes and fields. Life on the frontier was terrifying—and very dangerous.

Many settlers died, but mostly it was the Indians who were losing out. The white people's diseases, weapons, and numbers were too much to withstand. Anyone could see that the tribes were being destroyed. Some leaders, like George Washington and James Madison, tried to find ways to protect the Indians. It wasn't easy to do. Jedidiah Morse (who wrote the first American geography book) said that all Indians should live west of the Mississippi. Most white Americans—including Thomas Jefferson and Andrew Jackson—agreed. They all thought the Indians would be safe there, and could live in peace. (They couldn't see into the future and know that settlers would take that western land, too.)

The Cherokees didn't want to move. They loved their land.

Congress passed an Indian Removal Act (in 1830). That law made it legal for the president to move the tribes west. President Jackson was eager to do so. In his seventh Annual Message to Congress, he said the Indians would be moved to reserved lands west of the Mississippi. "The pledge of the United States has been given by Congress that the country destined for the residence of this people

John Ross (left), a Cherokee leader, resisted the resettlement of his people. But in the end they were forced to join the Cherokees led by chiefs such as George Lowrey (right), who had already agreed to the move to the Indian Territory.

> **Slavery seems** to have been part of Cherokee ways as far back as any could remember. At first the slaves came from enemy tribes; then a few slaves were whites; gradually more and more of them were blacks.

> **A song** some Georgians sang goes like this:
>
> All I ask in this creation
> Is a pretty little wife and a
>   big plantation
> Way up yonder
> In the Cherokee nation.

**117**

**There is** more to the Cherokee story than you'll find in this chapter. Much more. If you are interested you might want to do some research, especially about a Supreme Court case called *Cherokee Nation* v. *Georgia*.

shall be forever 'secured and guaranteed to them.'"

Meanwhile, gold was discovered on Cherokee land in Georgia, and gold hunters came by the thousands. They brought their guns with them. The Cherokees had no choice. Even though they had their own nation and were governing themselves, they had to go.

The Georgia government divided up the Cherokee land. They held a lottery and gave the land to white settlers. Soldiers helped those settlers move into beautiful Cherokee homes. They took the Cherokee farms and orchards. President Jackson said there was nothing he could do about it. The truth was, he didn't want to do anything about it.

By the time they had to remove westward, few Cherokees lived in the old ways as this man did, hunting, fishing, and gathering. Most were farmers and traders.

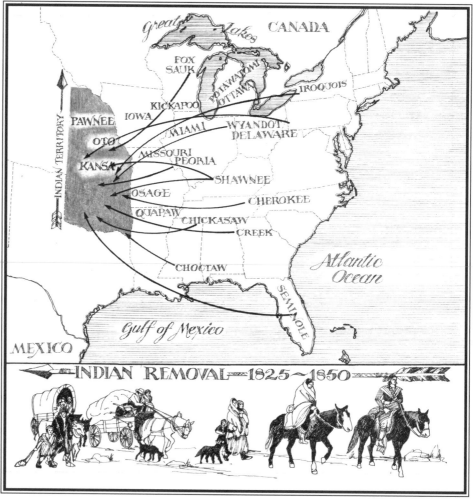

# 24 A Time to Weep

In 1832 Black Hawk and his son Whirling Thunder led a force of Sauk and Fox Indians in a fierce attempt to get back their homelands in Illinois.

It was called the Trail of Tears. And it was a trail, a long trail west, that people were forced to walk. As they went they wept, because they didn't want to go. They didn't want to leave their homes, their farms, their hunting grounds, the land of their fathers and mothers.

The people who wept were Native Americans. They were being forced to go by white soldiers with guns, and by a president, Andrew Jackson, who was famous as an Indian fighter.

The Cherokees were stubborn. They refused to move west. They refused to give up their land. They appealed to the government. Two congressmen, Henry Clay and Daniel Webster, said the Indians were right: their land was their land and no one else's. In 1832 Clay ran for president against Andrew Jackson (who was seeking a second term). The Cherokees prayed that Clay would win—everyone seemed to think he would—but he didn't. Jackson was popular. He was a man of the people, a man of the frontier, in many ways a good president. Most U.S. citizens agreed that the Indians should move west. They approved of Jackson's Indian policy. But sometimes the majority is not right.

The case of the Cherokees was argued before the Supreme Court. Court cases are named for those who are on opposite sides in the conflict. This case was called *Worcester* v. *Georgia*. The *v* is for *versus*, which is Latin and means "against," or "opposed to." It was Georgia (then the largest of the 27 states) against Worcester.

Worcester (say: WUSS-ter) was a man named Samuel Worcester. He

## Prey to Plunderers

*Evan Jones, a missionary, was a witness to the Cherokee exodus in 1838:*

The Cherokees are nearly all prisoners. They have been dragged from their houses, and encamped at the forts and military posts, all over the nation. In Georgia, especially, multitudes were allowed no time to take anything with them, except the clothes they had on. Well-furnished houses were left a prey to plunderers, who, like hungry wolves, follow in the train of the captors. These wretches rifle the houses, and strip the helpless, unoffending owners of all they have on earth. Females, who have habituated to comforts and comparative affluence, are driven on foot before the bayonets of brutal men.

When the Supreme Court ruled that the Cherokees had the right to live in their own nation, President Andrew Jackson—shown here as a "Great White Father," with Indians reduced to the status of dolls or puppets—said, "John Marshall has made his decision; now let him enforce it."

was a Congregational minister, a missionary, who had come to the Indian territory to teach school and to preach Christian doctrine.

Georgia passed a law saying all white men in the Indian portion of the state needed to be licensed. The law was really meant to get rid of the preachers who were taking the side of the Cherokees. No one would give Sam Worcester a license.

Worcester was arrested— twice. In order to stay out of jail, he moved to nearby North Carolina and continued his work from there. His wife and children stayed in Georgia. Then his baby died. When he came home to be with his wife Worcester was arrested again. He was tried, found guilty, and sentenced to four years in prison at hard labor. He appealed his case to the Supreme Court.

The old man who was chief justice of the Supreme Court could remember British times and the Indian treaties of the old days. He was a man with a mind straight as an Indian's arrow. Do you know his name? (Hint: he was Thomas Jefferson's cousin.) The great chief justice wrote a famous opinion in *Worcester* v. *Georgia*. (A justice's written words are called an *opinion*.) That opinion is still read and cited as a statement of fairness on human rights.

It went beyond the case of Samuel Worcester. The court ruled on the issue of Indian ownership of their own land and their right to govern themselves. Great Britain, said the chief justice,

> *considered [the Indians] as nations capable of maintaining the relations of peace and war; of governing themselves…and she made treaties with them, the obligation of which she acknowledged.*

Now here is the important part:

> *The Cherokee nation, then, is a distinct community, occupying its own territory…in which the laws of Georgia can have no force, and which the citizens of Georgia have no right to enter, but with the assent of the Cherokees themselves.*

The most famous of all chief justices (yes, it was John Marshall) had a whole lot more to say, but you get the idea. The Cherokees had won the right to their own land. (And to decide if Sam Worcester could teach and live on that land.) The Supreme Court said that *the Indians have a present right of possession.* In other words, it was unconstitutional to push the Indians from their land.

Only it didn't matter. The president—Andrew Jackson—refused to enforce the law. Our American system of checks and balances failed. I'm sorry to have to write this. It was a terrible moment in United States history. But the truth needs to be told.

The Cherokee leader Stand Watie resisted removal. The Union's abandonment of the Cherokees would backfire later: many, such as Watie, who led a regiment, fought for the South in the Civil War.

## Falling Trees—and Hills Bare of Bear

Frenchman François André Michaux (FRON-swah-ON-dray-me-SHO) spent years in America studying its trees and flowers. He is known as the "father of American forestry." Michaux was horrified by the waste of oak, hemlock, and cypress. American homes and factories and steamboats and railroads were burning the forests as fuel. Timber was scarce in the East, and so were the native grasses; they had been destroyed by overgrazing.

Farmers had to send to England for clover and timothy grass seed to sow for their cattle to eat. A few people, like Michaux, worried about nature and the environment, but most Americans thought the

land was endless, and its resources endless, too. They weren't, of course. Before the end of the 18th century, just about all the original forest east of the Appalachian Mountains had been cut and burned.

In 1800, Daniel Boone returned to Kentucky (he'd been away for 20 years) and was depressed by what he found. When he had lived there, he said, *you could not have walked more than a mile in any direction without shooting a Buck or a Bear. Thousands of Buffalo roamed the Kentucky hills and land that looked as if it never would become poor. But when I [returned]…a few signs only of Bear were to be seen. As to Deer I saw none.*

**The Supreme**

Court ruled that the Cherokees' land belonged to them. The Court also said that the Indians were not able to be entirely independent of the government. "They are in a state of pupilage; their relation to the United States resembles that of a ward to his guardian. They look to our government for protection; rely on its kindness and its power; appeal to it for relief of their wants; and address the President as their great father."

People in Georgia wanted Indian land. (I don't want to pick just on Georgia. The same thing happened in many other states.) And so the Indians of the eastern woodlands went west. Some fought before they went. The Sauk and Fox in Illinois fought especially hard, but their cause was hopeless.

The Choctaws were first; they moved in 1831. Three years later the Chickasaws trudged west. The Creeks signed a treaty that said "they shall be free to go or stay, as they please." It didn't matter. In 1836 they were sent west—some with chains around their necks.

The Cherokees set out in 1838. They left their homes and walked west, against their wishes. They went from their lush, fertile mountain lands to a region beyond the Mississippi that few people wanted

## Barron v. Baltimore

John Barron owned a wharf (a boat dock) in Baltimore harbor. It was a very profitable wharf, because it was on deep water and big clipper ships and cargo carriers could dock and unload there. Because of its fine harbor, Baltimore had grown large—bigger than the Maryland capital, Annapolis. That growth had led to a lot of construction. The city of Baltimore was careless with the dirt from all the building and roadmaking. Every time it rained, that dirt—often called silt—washed into the harbor. Most of it seemed to be settling right at Barron's Wharf.

Barron and 10 other wharf owners wrote to the city, saying: *For within one year last past, many parts of Our Wharves have been filled up with sand and dirt from four to six feet....if the City Commissioners continue digging ditches to alter the Water Course...Our Wharves will be useless to us and intirely ruined.* But the city of Baltimore paid no attention, so the dock owners went to court asking that the city dredge up the sand.

The Constitution's Fifth Amendment says that *private property shall not be taken for public use without just compensation*. (*Just* means fair. *Compensation* is payment.) The dock owners' property

Black Hawk and five other Sauk and Fox leaders wearing ball and chain after their rebellion and capture in 1832. Afterward, their tribes were forced to move west—again and again and again.

## Protected States' Rights—Not Citizens'

was being used as a public dump site. The Baltimore court said the dock owners should be compensated. But the city of Baltimore didn't want to pay. The city appealed the case, all the way to the Supreme Court. It was 1831, and this was Chief Justice John Marshall's last important constitutional case (and his first Bill of Rights case).

*A busy wharf similar to John Barron's, crowded with ships loading and unloading.*

He didn't seem to think it a difficult case at all. Marshall had created judicial review when he wrote these words: *It is... the duty of the judicial department to say what the law is.* What does this have to do with John Barron and his wharf? Well, as you just read, the Constitution (which is the highest law of the land) has words about private property and just compensation. They come from the Fifth Amendment, which is part of the Bill of Rights—and that is part of the Constitution; so the dock owners thought they were protected.

But, as you know, our nation was an experiment. We were trying to work out that federal system which divides power between the states and the central government. Justice Marshall said that the Constitution and its Bill of

Rights had nothing to do with the states (except in those parts that specifically deal with the states). This is what Justice Marshall said. *The constitution was ordained and established by the people of the United States for themselves, for their own government, and not for the government of the individual states.* Does that mean a state could take away your right to free speech? Yes, it does.

Remember, the founders worried about big government power. They didn't realize that state and local governments could be equally dangerous. (It was the state governments that were insisting on slavery.) John Marshall said that state constitutions and state bills of rights were meant to protect citi-

zens from state power. What if a state was unfair to a citizen? It was no business of the Supreme Court.

That idea did get changed. There was a war, a civil war, before it happened. Then three new amendments enlarged the Bill of Rights. One of them—it is the 14th Amendment—says: *No State shall make or enforce any law which shall...deprive any person of life, liberty, or property, without due process of law; nor deny to any person within its jurisdiction the equal protection of the laws.*

That came too late for John Barron, but today, the guarantees of the first 10 amendments are yours in federal, state, and local courts.

(at the time). They walked—the children, their parents, and the old people—on hot days and cold. They walked in rain and windstorm. Often there was not enough food; often there was no shelter. Always there was sadness, for one of every four of them died during the cruel march.

The government said the new land would be theirs forever. But when the white people moved west they forgot their promises to the Indians. They took their land again, and again, and again.

# 25 The Second Seminole War

**Osceola and the Seminoles refused to be made to leave their land.**

**This engraving of the "Dade Massacre" in 1836 was designed to terrify whites who were afraid of an alliance between slaves and Indians.**

*It may be regarded as certain that not a foot of land will ever be taken from the Indians without their consent. The sacredness of their rights is felt by all thinking persons in America.*

Thomas Jefferson wrote those words in 1786. It was now the 19th century, and land-seeking settlers—they were called homesteaders—had filled up the fine country in northern Florida. They were pushing their way south onto the land reserved for the Indians. Some white settlers were shooting and burning and looting. Some Seminoles were shooting and burning and looting. There was only one way to satisfy the homesteaders and protect the Indians, said most white people. The Seminoles must move.

The law was clear. The Indian Removal Act of 1830 stated that all Indians east of the Mississippi could be moved west to reserved Indian lands.

The Seminoles argued among themselves. Should they move? Could they resist? Seminole leaders went west to see the land the government wished to give them. It was then that they discovered something they could not accept: The United States government intended that they live with the White Stick Creeks! Many of the Seminoles had been Red Sticks. The White Sticks were their enemies. The White Sticks lived like white Americans: some even had plantations and slaves. The Seminoles lived an Indian life. They had black friends. Some were blacks themselves. It wouldn't work. The Seminoles decided to resist.

Wiley Thompson, the U.S. government agent in Florida, wrote a letter to Washington, D.C., telling of a "bold and dashing young chief …vehemently opposed to removal." The young chief was Osceola.

Thompson wrote another letter: "one of the most bold, daring, and intrepid chiefs in the nation…hostile to emigration…came to my office and insulted me." Again, he was writing of Osceola.

In the beginning, Osceola tried to reason with those who were invading his land. He said the Seminoles would agree to move west if the government would provide a special agent to protect his people from their Creek enemies. The government would not. After that, the Seminoles held a secret meeting and decided to resist all attempts to move them. Any Seminole who made plans to emigrate would be killed. The war was on. Osceola and his men appeared out of the swamps, attacked U.S. soldiers or settlers, and disappeared. Their raiding parties terrorized plantations and ambushed soldiers. They wrecked the sugarcane industry. They killed Wiley Thompson.

Major Francis L. Dade marched to Florida with two companies of troops. He had orders to conquer the Seminoles. Dade's troops were

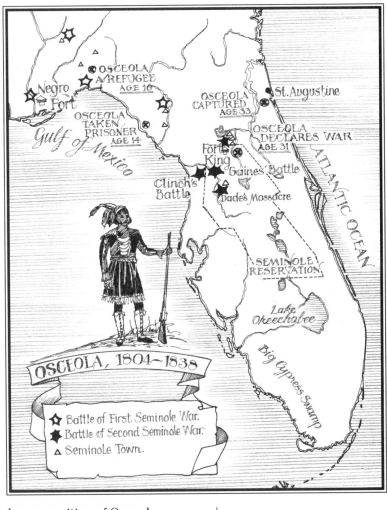

OSCEOLA, 1804~1838

☆ Battle of First Seminole War.
★ Battle of Second Seminole War.
△ Seminole Town.

## Impossible to Live Together

*Opothleyoholo, a Creek leader, speaks in 1835:*

Our people yet abhor the idea of leaving all that is dear to them—the graves of their relatives; but circumstances have changed their opinions; they have become convinced of their true situation; that they cannot live in the same field with the white man. Our people have done that which we did not believe they would have done at the time we made the treaty; they have sold their reservations—it is done and cannot now be helped; the white man has taken possession, and has every advantage over us; it is impossible for the red and white man to live together.

slaughtered. General Duncan Clinch was sent with a large American army—mainly to get Osceola. He didn't. General Edmund P. Gaines came with a still bigger force. He failed. General Winfield Scott assembled the largest army of all. Scott had a grand plan. He divided his soldiers into three battle lines to search out the Native Americans and crush them. But it was Scott's soldiers who got crushed.

General Robert Call was sure he could do better. He didn't. It was now 1836. The American soldiers were suffering from heat, disease, and frustration—as well as Indian attack. Bloodhounds—dogs from Cuba—were sent to sniff Indians out of the swamps. They didn't. General Thomas S. Jesup took over. His army was the largest yet. The Seminoles fought on. But Osceola had caught malaria; he and his men were tired.

By now, most Americans were sick of the war. It had taken 1,500 American soldiers' lives and cost more than $20 million. This war in the Florida swamps was much like a 20th-century war fought in the jungles of Vietnam. It pitted strong armies against small guerrilla bands that hit, ran, and humiliated their foe. The war became immensely unpopular with the American people.

Then one day, the Seminoles raised a white flag of truce. They wished to exchange prisoners. General Jesup had his troops secretly surround the conference area. At his signal, hidden men leapt out of the shrubs and overwhelmed the Indians. Osceola was a prisoner.

Osceola's capture under a white flag of truce outraged most American citizens. John Ross, a respected Cherokee peacemaker, wrote to the Secretary of War protesting "this unprecedented violation of that sacred rule...of treating with all due respect those who...presented themselves under a flag of truce."

An editorial in the *Charleston Courier* said of Osceola, "Treacherous he may have been, but...he was provoked by treachery and captured by treachery. We now owe him the respect which the brave ever feel toward the brave."

Colonel Ethan Allen Hitchcock wrote, "I came [to Florida] as a volunteer...to be of service in punishing as I thought, the Indians. I now come, with the persuasion that the Indians have been wronged."

But there was no other way to capture Osceola, whined General Jesup. "[Not a] single instance ever occurred of a first rate warrior having surrendered."

Osceola's malaria grew worse in prison. He asked that the other Indian chiefs, his two wives, and his two children be brought to his jail cell. It was January 30, 1838. A doctor who was there described what happened next to George Catlin, who had painted Osceola's portrait. Catlin wrote it down:

*He rose up in his bed…and put on his shirt, his leggings and his moccasins—girded on his war belt—his bullet pouch and powderhorn, and laid his knife by the side of him on the floor. He then called for his red paint, and his looking-glass, which was held before him, when he deliberately painted one-half of his face, his neck and throat—his wrists—the backs of his hands, and the handle of his knife, red with vermillion…His knife he then placed in its sheath, under his belt and he carefully arranged his turban on his head and his three ostrich plumes that he was accustomed to wearing in it. Being thus prepared in full dress…and with the most…pleasing smiles, extended his hand to me and to all the officers and chiefs that were around him; and shook hands with us all in dead silence; and also with his wives and little children; he made a signal for them to lower him down upon his bed, which was done, and he then slowly drew from his war-belt his scalping knife, which he firmly grasped in his right hand, laying it across the other on his breast, and a moment later smiled away his last breath, without a struggle or a groan.*

Most of the Seminoles who were left went west to the Indian territory. Some still refused to go. In 1842 the government gave up; about 300 Seminoles remained in Florida. In the end, no one won the Seminole War.

**Osceola showed what he thought of the removal treaty by skewering it with a knife. Even after his death, when most Seminoles had gone west, a few, like these hiding from the army in a mangrove swamp, held out.**

# 26 History's Paradox

Like whips, or a ball and chain, this bridle was used to punish or control slaves.

> **I have been in sorrow's kitchen and I licked the pots clean.**
> —Slave proverb

Here comes some difficult history. Put on your thinking cap—you're going to need it. The difficulty has to do with a paradox. A paradox is a contradiction. It may be good and bad at the same time. It is believing one thing and doing something else.

America was born of an idea and a dream. The dream was of a paradise: a land of unbelievable beauty. The idea was that this heavenly land should be a place of freedom and justice for all.

This is the paradox: America has been both dream and nightmare.

It is people who are the problem.

It is people who are the solution.

Some people have loved the land and cherished and conserved it.

Others have spoiled and ravished and destroyed.

Some Americans have lived on the land in freedom.

Other Americans have been slaves.

How has all this happened?

The first people came from Asia to a land of natural abundance—filled with flowers, animals, birds, and streams. But soon after men and women arrived, some creatures became extinct.

Norsemen and women landed, saw the beauty of the land, and killed natives sleeping under their canoes. That set a cruel pattern for the invading forces to come.

Columbus saw nothing wrong with taking the people he mistakenly named "Indians" and making them slaves. He was doing what many others did, without asking himself if it was right or wrong.

And the Founders—who wrote of liberty and equality in a remarkable declaration and a splendid constitution—gave their consent to

> **If they [white servants] get a bad master, they give warning and go hive to another. They have their liberty. That's just what we want. We don't mind hard work, night and day, sick or well...[but] we must not speak up nor look amiss, however much we be abused.**
> —Mary Prince, 1790s slave

| DISTRICTS. | Free white Males of sixteen years and upwards, including Heads of Families. | Free white Males under sixteen years. | Free white Females, including heads of families. | All other free Persons. | Slaves. | Total. |
|---|---|---|---|---|---|---|
| Vermont, | 22,435 | 22,328 | 40,505 | 252 | 16 | 85,539 |
| New-Hampshire, | 36,086 | 34,851 | 70,160 | 630 | 158 | 141,885 |
| Maine, | 24,384 | 24,748 | 46,870 | 538 | none | 96,540 |
| Massachusetts, | 95,453 | 87,289 | 190,582 | 5,463 | none | 378,787 |
| Rhode-Island, | 16,019 | 15,799 | 32,652 | 3,407 | 948 | 68,825 |
| Connecticut, | 60,523 | 54,403 | 117,448 | 2,808 | 2,764 | 237,946 |
| New-York, | 83,700 | 78,122 | 152,320 | 4,654 | 21,324 | 340,120 |
| New-Jersey, | 45,251 | 41,416 | 83,287 | 2,762 | 11,453 | 184,139 |
| Pennsylvania, | 110,788 | 106,948 | 206,363 | 6,537 | 3,737 | 434,373 |
| Delaware, | 11,783 | 12,143 | 22,384 | 3,899 | 8,887 | 59,094 |
| Maryland, | 55,915 | 51,339 | 101,395 | 8,043 | 103,036 | 319,728 |
| Virginia, | 110,936 | 116,135 | 215,046 | 12,866 | 292,627 | 747,610 |
| Kentucky, | 15,154 | 17,057 | 28,922 | 114 | 12,430 | 73,677 |
| North-Carolina, | 69,988 | 77,506 | 140,710 | 4,975 | 100,572 | 393,751 |
| South-Carolina, | 120,000 | | | 100,000 | 220,000 | |
| Georgia, | 13,103 | 14,044 | 25,739 | 398 | 29,264 | 82,548 |
| | | | | | 687,216 | 3,641,602 |

In this summary of the 1790 census the fifth column shows the number of slaves per state. "All other persons" means free blacks.

| | S. CAROLINA | N. CAROLINA |
|---|---|---|
| Free white Males of twenty-years and upwards, including heads of Families. | 13,103 | 69,988 |
| Free Males under twenty-one years of age. | 14,044 | 77,506 |
| Free white Females, including heads of families. | 25,739 | 140,710 |
| All other free Persons. | 398 | 4,975 |
| Slaves. | 29,264 | 100,572 |

The Constitution said that a state could have one representative for every 40,000 people. A slave counted as three-fifths of a person, even though no slave could vote—which gave the slave states an advantage.

***Hypocrisy*** is saying one thing when you believe another.

slavery, which was the very opposite of liberty and equality.

How come? Were all these people liars? Or dummies?

Of course they weren't. They were trying hard to do the best they could in a world where ideals weren't all that mattered. There were also powerful forces called selfishness, bigotry, hypocrisy, and cruelty.

Which brings us to the 19th century in America and that awful paradox: slavery in the land of the free.

The fight to end the paradox, to get rid of the horror of slavery, was to be the most important battle in all our history. It led to war and, finally, to the end of slavery.

There is more to this paradox than you may know. (Paradoxes are like that.) Some blacks, who were free, owned slaves. In 1830, 1,556 free black masters in eight Southern states owned 7,188 slaves. Are you surprised that black people owned slaves? Blacks owned slaves for the same reasons that whites did. It made them prosperous. The economic system in the South made it profitable to own slaves. Of course, that didn't make it right.

After 1808 it was against the law to bring Africans into the United States as slaves. In 1820 the penalty for breaking that law became death. Still, between 1820 and 1850, the number of slaves

## 1 Wench Nam. Eve & Child

*If you were a slave, you could be bought and sold just as if you were an animal or a piece of furniture. In 1811 Abraham van Vleek bought a collection of goods from Barent van Dupail (who was probably selling because of financial problems). Far and away the most expensive item on the list was a slave named Eve and her child.*

*1 Faning Mill..............$ 17.25*

1 Red face Cow...........$ 13.25
1 Yearling Calf.................4.25
1 Plough............................1.6
1 Wench Nam. Eve
  & Child.....................156.00
8 Fancy Chairs................9.25
1 Looking Glass. 6 Silver Table Spoons. 6 So. Tea Spoons 10 China saucers. 11 Wo. Cups 1 Tea Pot. Sugar and Milk Cups. 3 Plates Dish
and tea Bord................. 35.12½
[Total]....................$236.18½

129

African Americans mingled Old and New World styles in a new way of life. The dress of this fruit and vegetable seller in a Southern market city is African in tradition.

doubled: from a million and half to more than three million. Many of that number were children born in the United States, but many others were new slaves brought in illegally.

Some people were lawbreakers, stealing blacks from their African homes and slipping them into the country. The profits in slave trading were huge; some people thought it worth the risk. (Like drug dealers, the people who made the money didn't care if they were destroying other people's lives.)

But a few people—whites and free blacks—were doing everything possible to end the slave trade and to end slavery itself. Some thought the way to end slavery was to return blacks to Africa. Paul Cuffe went to Africa to start a colony for black Americans who wished to return to Africa. James Madison was president of the African Colonization Society; so was Senator Henry Clay. Many well-meaning people joined the colonization movement. They had the idea that black and white people couldn't live together on equal terms, so the only path to peace was separation. What do you think of that idea?

Most blacks who lived in America didn't want to go to Africa. They found they had little in common with Africans. They were Americans, and they wanted to be free and have all the same rights as other Americans.

In an article in a Maryland newspaper a black writer said:

> Though our bodies differ in color from yours; yet our souls are similar in a desire for freedom. [Difference] in color…can never constitute a [difference] in rights. Reason is shocked at the absurdity! Humanity revolts at the idea!…Why then are we held in slavery?…Ye fathers of your country; friends of liberty and of mankind, behold our chains!…To you we look up for justice—deny it not—it is our right.

And yet, for most, justice was denied.

In the Declaration of Independence, the Founders told us that we were all created equal and that we all have the right to "life, liberty and the pursuit of happiness." No country before had goals like those. And

## 100 Dollars Reward.

RAN away in July 1847, a negro man by the name of BUCK. He is yellow complexion, about five feet six inches high, rather bow-legged, very quick in his movements, and when spoken to very slow to answer. He was in the possession of Robert F. Morris, at Hillsborough, when he went away, and is very likely still in that neighborhood; yet he was raised in Granville county, by Mrs. Blacknall, in the neighborhood of Winton, and may be in that neighborhood now. The above reward will be given for his apprehension, and delivery to me; or confinement in any jail so that I can get him.

WM. J. HAMLETT.

Mount Tirza, Person, N. C., Jan. 29.    —66

we had them in writing. Our Founders made fairness a national creed.

It doesn't sound very complicated—providing everyone with freedom, equality, and a chance to pursue happiness in a good land—but no nation has ever been completely fair to all its citizens.

Fairness is something you have to keep working at. Each generation has to do its job. In the 19th century, America's black people, along with fair-minded whites, would struggle and fight to end the paradox of slavery in a free nation. They understood that no one is free in a land where some are enslaved.

Slaves who escaped successfully cost their owners a lot of money. Besides that, every escaped slave gave hope and courage to the whole slave community. But many, like the man trapped in a tree above, didn't make it.

# 27 A Man Who Didn't Do as His Neighbors Did

**After the** Revolution, the Anglican Church in America became the Episcopal Church. In New England, most people became members of the Congregational Church.

It was a fine life for a young man like Robert Carter III. At one ball held at Nomini Hall when Robert was 15, he and his brother danced the night away and never went to bed at all.

**The Northern** Neck is bounded by the Potomac and Rappahannock rivers, with streams and estuaries and a sky filled with ducks, swans, and geese. The Lees lived there, and George Washington's brother, and so many leading families that it was once called the "Athens of America." (What does that mean? Why Athens? Where is Athens?)

Everyone agreed that Robert Carter III was different from most other people, and, in 1791, he proved it. He did something astonishing. What did he do? Hold on, I'll get to that soon. How was he different? To begin, Robert Carter was very, very, very rich.

He had 16 plantations, his own ships, an iron foundry, flour mills, textile workshops, and more than 500 slaves. Carter had a mansion in Williamsburg and a huge plantation home—Nomini (NOM-in-eye) Hall—on Virginia's prosperous, fertile Northern Neck peninsula. Nomini Hall sat on a high piece of land and could be seen six miles away.

Life was mighty fine for Robert Carter III. Ships stopped in his front yard; his 17 children had their own school and their own schoolmaster; and he and his aristocratic neighbors entertained at grand parties and balls. The Carters' dinners, every day, were like our Christmas and Thanksgiving dinners combined.

But, as I said, Robert Carter III wasn't like his other wealthy neighbors. He wasn't like his famous grandfather either. Robert Carter I was known as "King Carter," or "King Robin," and was the wealthiest of all the Virginians. When he died he was buried with pomp and honors. His tombstone said he was "possessed of ample wealth, blamelessly acquired." But those who knew him weren't fooled. Blameless? Hah! King Carter was a ruthless and greedy land grabber. (Soon after King Carter died, someone chalked on the tombstone: *Here lies Robin, but not*

*Robin Hood,/Here lies Robin, that never was good,/Here lies Robin, that God has forsaken,/Here lies Robin, the Devil has taken.)*

His grandson Robert Carter III wasn't satisfied with riches. He wanted something more. He wanted to understand the meaning of life. Robert Carter III set out to learn the truth about God and religion.

He lived at a good time for someone who wanted to undertake that search. It was the end of the 18th century and the beginning of the 19th. It was the Age of Reason. People were using their minds, reading books, and asking questions. Robert Carter had his own fine library, and when he wanted to talk to someone, it might be Thomas Jefferson, or George Mason, or James Madison.

It was his search for the truth about religion that made him different from most of his neighbors. Carter began life as an Anglican—it was England's church. According to Virginia's law, everyone had to go to that church at least once a month—or pay a fine. The Anglican Church was called the "established church." Other churches were "dissenting churches." Everyone paid taxes to support the Anglican Church, whether they believed in its teachings or not.

But laws can't control thought. Robert Carter, like many of the thinkers of his time, began looking for God in nature and in the remarkable order of the world. He became a deist (as did Jefferson, Washington, and Adams).

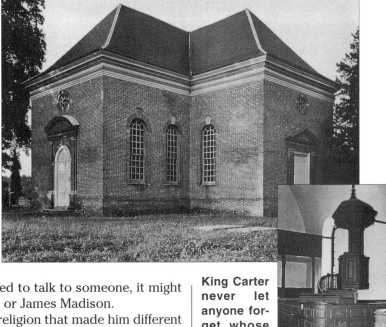

King Carter never let anyone forget whose money paid for Christ Church (above and inset). On Sundays the congregation didn't enter the church until the Carter coach had arrived. Christ Church, on Virginia's Northern Neck, is still one of the loveliest churches in America. Robert Carter I is buried in the churchyard.

Hannah Harris was one of Robert Carter's slaves. In 1792, knowing she would be freed the next year, she wrote this note to her owner. She asked to be allowed to buy the loom she wove on so she would have the means to support herself as a free woman.

**We know** quite a bit about the Carters because Robert Carter III hired a schoolmaster, Philip Fithian, who kept a diary. Fithian's *Journal* is fun to read. It tells about the way wealthy planters lived.

**Between 1782** and 1806 there were no laws or restrictions in Virginia on freeing slaves. Jefferson helped write the law that allowed people to free their slaves. (There was a kindly provision to the law that prevented slave owners from dumping elderly or sick slaves.) After 1806, new laws were passed that made it difficult to free slaves. Still, some people did it. In 1833, Virginia's John Randolph freed 383 slaves and bought farm equipment and land in a free state for them.

**"My mother** and myself begged Mr. Carter not to sell this child out of Fredg [plantation], he gave us his word and honor that he would not but as soon as we left him he sold the child."
—James Carter, slave of Landon Carter (uncle of Robert Carter III), telling of his eight-year-old sister, Judy. James never saw his sister again.

But deism didn't seem to satisfy him. Carter heard every preacher he could find. Sometimes he heard four or five in one week.

He became a Baptist. That surely astounded his neighbors. The Baptists, mostly, were poor folk. Many slaves were Baptists. Some of Carter's neighbors must have made fun of the Baptists.

But although Carter was a Baptist, he didn't stop his search. He listened to Methodist preachers. Both the Baptists and Methodists believed people were equal before God. Thomas Jefferson wrote that "all men are created equal." That was an idea that must have set Robert Carter thinking. No one treated him as an equal. With all his money and power he was used to special privileges. How would you act if you were Robert Carter?

Well, he didn't just sit around and act rich. He accomplished things. He got 136 citizens to sign a petition in support of the Virginia Statute for Religious Freedom. That bill, written by Thomas Jefferson, let Virginians choose their religion. They no longer had to pay taxes to support a church. They didn't even have to go to church if they didn't want to.

But it was what Robert Carter did in 1791 that really astonished and shocked everyone. Here are his own words telling about it:

> *Whereas I Robert Carter of Nomini Hall in the County of West-moreland & Commonwealth of Virginia [own]…many negroes & mulatto slaves…and Whereas I have for some time past been convinced that to retain them in Slavery is contrary to the true principles of Religion and justice….I do hereby declare that such…shall be emancipated.*

Emancipated! That means free! Robert Carter freed all his slaves! He did what his religious beliefs told him was right to do. What do you think his friends said? We don't know that, but we do know that some of his children tried to stop him. (Carter was one of the largest slave

## The Golden Rule

*This is a miniature portrait of Henry Laurens. What is the golden rule? See if you can find out.*

"You know, my dear son," wrote Henry Laurens to his son John, "I abhor slavery." *Abhor* means hate. Henry Laurens hated slavery. But he was a slave owner! He went on in his letter to explain: *I was born in a country where slavery had been established by British kings and parliaments as well as by the laws of that country ages before my existence...and I nevertheless disliked it.*

It was 1776, and Laurens—a wealthy South Carolina planter and merchant—was in Philadelphia as a member of the Continental Congress. He told John that he was: *not one of those who...wish to continue in slavery thousands who are as well entitled to freedom as themselves....the day I hope is approaching when, from principles of gratitude as well as justice, every man will strive to be foremost in showing his readiness to comply with the golden rule.*

Laurens made plans to free his slaves. He knew, he said, that he would *appear to many as a promoter not only of strange, but of dangerous doctrines*; still, he wrote, *I will do as much as I can in my time, and leave the rest to a better hand.*

When the Revolutionary War began, Henry Laurens went to Europe to try to raise money for the revolution. But he was captured by the British and put in the Tower of London (an infamous prison). Laurens was a prisoner until war's end, when he was sent home in exchange for Britain's general, Lord Cornwallis.

Those 15 months in the Tower ruined Henry Laurens's health. If his health had been good he almost surely would have been a delegate to the convention that wrote the Constitution. If he had represented South Carolina at the Constitutional Convention, do you think United States history might be different?

Laurens was said to be the wealthiest merchant in Charleston. That means he was very, very rich. He made his fortune selling rice, indigo, deerskins and—slaves. Does that change your opinion of his letter? And what about that phrase "leave the rest to a better hand"? What do you think that means?

Henry Laurens's famous letter to his son was purchased by a Boston political club as a gift for President Abraham Lincoln. It was being put into a special binding on the day Lincoln went to see a play at Ford's Theater. (More about that in Book 6 of *A History of US*.)

owners in the United States.)

Americans were struggling with the problem of slavery. Actually, it was slavery that had made people aware of the importance of freedom. Jefferson proposed a constitution for Virginia that would gradually free the slaves (it wasn't adopted). Most of the men at the Constitutional Convention in Philadelphia said they hated slavery, but they didn't prohibit it. They didn't think they could do that and have a union of 13 states. That was a terrible tragedy for the whole nation.

Robert Carter decided for himself about slavery, but he wasn't finished with his search for truth. He left Nomini Hall (you couldn't run a plantation without slaves) and moved to Baltimore. There he joined another church. He was obsessed with the ideas of religion. Some people talked about the conflict between head and heart. Robert Carter was trying to find a balance between them: between reason and faith. Did he find all the answers he sought? Probably not. Perhaps no one ever does. But what do his actions tell you about the way he lived his life?

**Mary Stith** owned a small house in the middle of Williamsburg. When she died in 1813 she left her house and lot, as well as much of her estate (what she owned), to the "coloured people" who had served her. Her father, William Stith, had been a president of the College of William and Mary.

# 28 African Americans

This North Carolina slave cabin may have housed as many as 15 people at once.

Geologists tell us that in long-ago times, before there were people, the continents were attached. Africa and the Americas were part of one land until a violent wrenching of the earth tore them apart.

A few million years later, some of Africa's children were torn from their roots and transplanted to America.

Like the people who came from Europe and Asia, they were changed by the American land. Soon, very soon, they were different from the brothers and sisters they had left behind. Soon they were no longer Africans. Now they were Americans—African Americans.

An African may have come to the newly discovered land on the first of Columbus's voyages. By 1501, Africans were living in the Caribbean. In 1619 they were living in Jamestown, Virginia, and, a few years later, at Plymouth, Massachusetts.

Africans cleared the woods, tilled the soil, planted tobacco, and harvested cotton. They were farmers, trailblazers, mountain men, cowhands, and pioneers. They were Americans. They panned for gold, dug canals, and helped build railroads. They fought at Concord, Bunker Hill, and at Yorktown. A few were free, but most were slaves. Like all Americans, they longed for liberty; for their country and for themselves.

Those who could, spoke out. Elizabeth Freeman was a slave in Massachusetts when the Revolutionary War ended. Everywhere she

Elizabeth Freeman's portrait, done by the daughter of the lawyer who helped her win her freedom.

heard people talking about freedom and equality. She heard of the Declaration of Independence. Why shouldn't she be free, she wondered?

Elizabeth Freeman went to a lawyer and asked him to help her. In 1781 her case was heard in a court in Great Barrington, Massachusetts. The jury agreed with Elizabeth Freeman, and she was freed.

In 1783, Nathaniel Jennison beat up Quock Walker. Walker went to court to protest. Jennison told the judge he had every right to beat Walker because Walker was his slave. The judge, William Cushing, chief justice of the Massachusetts Superior Court, didn't agree. He said that:

> As to the doctrine of slavery and the right...to hold Africans...and sell and treat them as...horses and cattle...whatever [people have believed before]...a different idea has taken place with the people of America, more favorable to the natural rights of mankind....The idea of slavery is inconsistent with our...constitution.

Quock Walker was a free man.

Paul Cuffe's father was a free black who had been born a slave. His mother was a Wampanoag Indian. Cuffe went to sea, worked hard, and became the rich owner of a fleet of ships. But, because he was black, he wasn't allowed to vote in Massachusetts. Cuffe refused to pay his taxes. He said, "No taxation without representation." He appealed to the Massachusetts court, reminding the court that blacks and Indians had fought in the Revolutionary War. Cuffe lost his case, but the Massachusetts legislature then passed a law giving black people the same rights as whites. Cuffe had won the right to vote.

**Paul Cuffe's fellow Quakers buried him in a segregated part of their cemetery.**

James Forten served as a powder boy on a ship during the Revolutionary War. After the war, he invented a sailmaking device and made a fortune. Forten gave much of his money to help the cause of black freedom. "The spirit of freedom is marching with rapid strides and causing tyrants to tremble," wrote Forten. "May America awake..."

Lemuel Haynes was a soldier in the Continental army. He fought as a minuteman at the Battle of Lexington. After the war, in 1785, he became a minister in the Congregational Church. That was unusual. You see, Haynes's father was a slave, his mother was a white woman, and he was raised by foster parents. But he didn't let obstacles get in his way. Lemuel Haynes had a splendid mind and a sense of humor that

## Torn Apart

My brothers and sisters were bid off first, and one by one, while my mother, paralyzed with grief, held me by the hand. Then I was offered....My mother...pushed through the crowd while the bidding for me was going on, to the spot where Riley was standing. She fell at his feet, and clung to his knees, entreating [begging] him...to buy her baby [Josiah] as well as herself, and spare to her at least one of her little ones.

*This was written by Josiah Henson, a former slave, who escaped to Canada, became a Christian minister, and wrote his autobiography. The auction took place in 1795. Josiah was five. Isaac Riley did not buy Josiah's mother. Instead he kicked her, and left her weeping and childless.*

## On Slavery

*George R. Allen, a student at the New York African Free School, was 12 in 1828 when he wrote this poem:*

**S**lavery, oh, thou cruel stain,
Thou dost fill my heart with pain:
See my brother, there he stands,
Chained by slavery's cruel bands.

**C**ould we not feel a brother's woes,
Relieve the want he undergoes,
Snatch him from slavery's cruel smart,
And to him freedom's joy impart?

**To those** who suggested that free blacks should return to Africa, Richard Allen replied, "This land which we have watered with our tears is now our mother country."

When yellow fever struck Philadelphia, Absalom Jones stayed in the city to nurse the sick and bury the dead.

Lemuel Haynes was pastor of a mostly white church. Here, he gives one of his famous sermons. The scene was painted on a tray.

left people chuckling. When he became pastor of a church in West Rutland, Vermont, the church had 42 members; when he left, 30 years later, there were more than 300.

Maybe he attracted people because he had something to say. Some of his sermons were printed, and even read in England. In 1804, Middlebury College awarded an honorary degree to Lemuel Haynes. (Colleges give honorary degrees to outstanding people.)

Richard Allen—a bricklayer, a preacher, and a slave—converted his master to the Methodist religion. The master allowed Allen to work extra hours and buy his freedom. One Sunday morning, Allen and his friend Absalom Jones were in a Philadelphia church, on their knees, praying. Some white men rudely grabbed them and said that because Allen and Jones were black, they would have to sit upstairs.

Allen and Jones walked out of the church. Richard Allen founded the African Methodist Episcopal Church and became its first bishop. Then he opened a school for black children in Philadelphia. Allen and Jones founded the Free

Free blacks, Bishop Richard Allen said, "should help forward the cause of freedom."

African Society to fight against slavery. That wasn't all they did. In 1812, when the British burned Washington, Jones and Allen got 2,500 black troops to help protect Philadelphia. (Why weren't they needed? Do you remember what happened?)

These blacks were not alone—there were whites who cared. There had been since the earliest days of the colonies. In 1700, a Puritan judge, Samuel Sewall of Massachusetts, wrote, "Liberty is in real value next unto life: None ought to part with it themselves, or deprive others of it." The "others" he was talking about were blacks.

John Woolman, a white New Jersey Quaker, wrote in 1763, "I believe…a heavy account lies against us as a civil society for oppressions combated against people who did not injure us." He, too, was talking about blacks and slavery.

Sometimes the eye of an outsider sees a scene more vividly than those at home. In 1812 a Russian visitor painted these frenzied revivalist Methodists dancing and whirling to banish the devil in a Philadelphia alley.

So was James Madison when he wrote, "The magnitude of this evil among us is so deeply felt…that no merit could be greater than that of devising a satisfactory remedy for it."

Thomas Jefferson wrote, "Indeed I tremble for my country when I reflect that God is just." And the Marquis de Lafayette said, "I never would have drawn my sword in the cause of America if I could have conceived that thereby I was helping to found a nation of slaves." President John Quincy Adams called slavery "a cancer gnawing at America."

With all those brilliant people speaking out against slavery, you would think it would be easy to end it. It wasn't. Slaves represented money, and most people don't like to part with their money. Still, a few blacks were becoming free: some were running away, some were earning their freedom, some were freed by their owners.

## A Baptist's Conscience Pricked

*From Baptist leader John Mason Peck's journal, entry for January 1, 1842.*

Today I attended for a few moments a sale in the [Nashville] market-place. A negro boy was sold who appeared about twelve years old. He stood by the auctioneer on the market-bench with his hat off, crying and sobbing, his countenance a picture of woe. I know not the circumstances; but it was the first human being I ever saw set up for sale, and it filled me with indescribable emotions. Slavery in Tennessee is certainly not as oppressive, inhuman and depressing as the state of the poorer classes of society in England, Ireland, and many ports of Continental Europe; yet slavery in its best state is a violation of man's nature and of the Christian law of love.

# 29 The King and His People

A coffle is a walking jail made by shackling people with wrist and leg irons and chaining them together. Slaves ate and slept in the chains.

It was a king who was messing things up. The Founders had warned about kings. But they were thinking of those fellows on the English throne. It never occurred to them that tyrannical monarchs might take other forms.

This king was called Cotton. King Cotton sat on his throne for about 60 years: from the time Eli Whitney invented the cotton gin until the terrible war between the states.

Cotton's throne was built of the arms, legs, backs, and hearts of Americans—black Americans—who did all the hard work in the king's empire.

The South that King Cotton ruled was different from the South of the 18th-century tobacco and rice planters. King Cotton's South was a new South, with new lands and new names. It grew fast, very fast, from the frontier lands of the former colonies, to the new states of Mississippi and Louisiana, and on to Texas. Wherever land was cheap and cotton would grow there were men eager to make their fortunes.

Most of the newly rich men—the princes that King Cotton crowned —were frontier boys, born in log cabins. These new plantation owners imitated the style of the Virginia and South Carolina aristocrats, but had the energy and ruthlessness of the backwoods. Andrew Jackson was one of them, John C. Calhoun another, and Jefferson Davis yet another.

They were smart, and lucky too. Each of them bought a little land, worked it, bought a slave or two, worked them, bought more land, and soon had a whole plantation.

There weren't many big plantation owners, but they set the tone for the South. Most of the population was made up of yeoman farmers: people who owned small farms, as northern farmers did. Most yeomen

**The cotton** gin was invented in 1793. The Civil War—also called the War Between the States—began in 1861.

**Andrew Jackson,** as you know, became president; John Calhoun was an important senator; and Jefferson Davis was president of the Confederate states—the South—during the Civil War.

Some dangerous jobs were done by white people; slaves were too valuable to risk damaging them.

didn't have slaves, though a few owned a slave or two. In addition, there were poor whites, who had enough to eat and not much more.

At the bottom of the ladder were the slaves, who were owned by people who talked of liberty.

It was a slow-paced region, where crops grew easily. And that made the South a leisurely place. Perhaps because of that people had time to be polite and kindly—and they were. They had time to tell stories. Southerners were among the best storytellers in the nation. Some say it was the Southern land that made people tell stories—a land of dreamy mists and fogs and mysterious forests where Spanish moss hangs on the trees like lacy gray curtains. Walk in woods like that and you may see goblins and spooks and little people; soon you'll think up your own stories.

Black Southerners told some stories that they brought from their African homelands. And others—like those of Brer Rabbit and Brer Fox—were made up here.

There is something else about the Old South that I need to mention. It had a tradition of violence: duels were frequent and so were lynchings. In the 20 years from 1840 to 1860, 300 people were lynched by mobs in Southern states. (To lynch means to kill without a fair trial.) Most of those lynched were white. There was almost no lynching of slaves—they were worth money.

If you want to understand about slavery and the conflict between blacks and whites, you will need to do a lot of reading and thinking. Especially if you want to be fair.

**This was** a slave song:
We raise the wheat,
They give us the corn;
We bake the bread,
They give us the crust;
We sift the meal,
They give us the husk;
We peel the meat,
They give us the skin;
And that's the way
They take us in.

**Some words** from another slave song:
Rabbit in the briar patch,
Squirrel in the tree,
Wish I could go hunting,
But I ain't free.
The big bee flies high,
The little bee makes the honey.
The black folks make the cotton
And the white folks get the money.

The word **lynch** came into English about 1830, after John Lynch, a Virginia justice of the peace, set up his own trials and executed people without *due process of law*.

"Simon has permission to sell Turkeys, Eggs &c." A slave had to have a permission slip to travel or sell goods in the market.

*Simon has permission to sell Turkeys. Eggs*
*Feb—14th 1854*

141

**At a slave auction, husbands could be sold away from their wives or mothers from babies.**

One day my mammy did something and old master...took a leather strap and whooped her. I remember that I started crying.
—OLIVER BELL, REMEMBERING SLAVERY IN ALABAMA.

You will read tales of kind masters and happy slaves and you will read stories of viciousness, cruelty, and abuse. There is both truth and exaggeration in most of those stories.

Yes, there was terrible, brutish, inexcusable meanness in slavery. But most slave owners—even if they were cruel—thought of their slaves as valuable property. They might beat them, but they tried not to do them serious harm. They needed to keep their property healthy.

Yes, there were kind slave owners who cared for the people they owned and treated them well. There was genuine affection between blacks and whites. But how would you like to be owned, even by someone nice? And always, behind the sweet-smelling magnolia trees and the white-columned mansions, there were chains, whips, and guns. There would have been no slavery without them.

Remember, in the time of George Washington most Southern whites hadn't liked slavery; they just didn't know how to end a bad system. As time went on, Southerners tried to tell each other that slavery was all right. They said that God had created some people to be slaves. They said black people weren't as smart as white people. Then, to make that true, they passed laws that said it was a crime to teach any black person to read and write. One white woman in Norfolk, Virginia, who taught some free blacks in her home, was arrested and spent a month in jail. Can you see what was happening? Whites were losing their freedom, too.

## Ending the Slave Trade

Article I, Section 9 of the Constitution said that the slave trade could not be prohibited by Congress until 1808. President Thomas Jefferson knew that time would soon arrive. Would Congress pass laws to end the trade in human beings? In his annual presidential message to Congress and the people, Jefferson said:
*I congratulate you, fellow citizens, on the approach of the* period at which you may interpose your authority constitutionally to withdraw the citizens of the United States from all further participation in those violations of human rights which have been so long continued on the unoffending inhabitants of Africa. That means that citizens could now demand that Congress introduce a bill ending the slave trade. On January 1, 1808, it was done.

The slaves' battle for liberty is the next act in the drama that began with the Declaration of Independence and those noble words *all men are created equal*. It will help free all Americans. The curtain won't come down with the end of slavery. The freedom drama will continue. Women will demand an act, and Native Americans. Can you think of others who will take roles as freedom fighters? Is there still a part for you to play?

# 30 Abolitionists Want to End Slavery

The famous plea on this abolitionist pottery medallion cries: "Am I not a man and a brother?"

*Abolition* (ab-uh-LISH-un)! Back in 1765 Americans had shouted the word. To *abolish* means to end or to do away with something. Before the Revolution it was the hated British stamp tax the American colonists wanted to abolish.

Then the word began to be used with a new meaning. It was the slave trade some wanted to abolish, and then slavery itself. In 1773, Ben Franklin wrote in a letter that "A disposition to abolish slavery prevails in North America." That—in plain English—means people in North America want to find a way to end slavery. Two years later Franklin helped found the American Abolition Society.

The official African slave trade did end—when the Constitution said it would—in 1808. But slavery continued. And an illegal slave trade began.

The problem was one of finding workers for jobs nobody wanted. No one had anything good to say about slavery—at least not in George Washington's day. But then, slowly, some people's ideas changed.

Partly it was because, during the 1820s and 1830s, some slaves rebelled and killed white people. After that, white Southerners started to be afraid of the slaves. Slavery became even more cruel. New laws were passed that gave slaves almost no rights at all. Some Southerners began finding excuses for slavery. Others began to say it was a fine way of life—for slave and master.

If you read the diaries and letters of white Southerners you will see there were many who knew better. Robert E. Lee, who was to become the South's most famous general in the Civil War, wrote in a letter, "Slavery is a moral evil in any society...more so to the white than to the

**Fear was** always a part of slavery. When you abuse people you are likely to be afraid of their anger. Thomas Jefferson called slavery a "wolf." He wrote, "We have the wolf by the ears, and we can neither hold him nor safely let him go." What did he mean by that?

**Nat Turner,** Denmark Vesey, and Gabriel Prosser were slaves who led freedom rebellions. None was successful.

**America in** the first half of the 19th century was like two separate nations. The North was becoming a modern, industrial, urban nation, with railroads, canals, steamboats, and factories—as well as farms. Many of the workers in the new factories and mills were immigrants. The South remained an agricultural region. Mostly, immigrants stayed away from the South.

**Moral** has to do with understanding the difference between good and evil. A person who is moral acts with goodness in mind. A person who is **immoral** doesn't care about the difference between right and wrong. What did Robert E. Lee mean? Why was slavery more of a moral evil for whites than for blacks?

black." But some people didn't care about morality. The South was having economic problems—and some Southern political leaders began blaming the North for those problems.

Virginia and South Carolina had once been very rich colonies. They had been the envy of all the Americas. Now the "Old South" was in decline. The tobacco land was worn out. Economic power—money—had moved to the new cotton states and to the new industries in the North.

The Southern leaders didn't seem to understand what was happening. Immigrants and ideas and inventions were changing the North. The South was left out of that excitement. Newcomers didn't want to move south. They knew that as workers they couldn't compete with slaves for jobs. The South became isolated; it didn't grow with the 19th century.

Southerners began to live in a world of olden times. They read stories of the old days and believed those days were better. They wouldn't admit they were trapped in an evil system that got worse and worse each year.

Religious groups got involved. In the North, the Quakers were at the center of the abolitionist movement. But the leaders of some other religions—North and South—defended slavery.

The South became very jealous of political power. So did the North. Each wanted to dominate the nation. Each was afraid of the other. Tempers flared. It got more intense than that old conflict between Federalists and Republicans. But, as long as Congress was divided evenly between slave states and free states, there was some stability.

Then, in 1820, Missouri asked to enter the Union as a slave state. Northerners were alarmed. If Missouri became a state, the North would be outvoted in Congress. Northern congressmen were afraid of what might happen next. Suppose Congress voted to allow slavery in all the states! The situation was tense. Finally, a solution was found. Here it is: Maine was carved from Massachusetts and made into a state, a free state. That kept the balance of free and slave states. That action was called the Missouri Compromise. The

## Willing Sacrifice

From the beginnings of slavery, blacks had fought enslavement. Some had run away to freedom. Some had joined Indian tribes. In 1804, a slave who took part in a rebellion was brought to trial. This is what he told the court: *I have nothing more to offer than what George Washington would have had to offer had he been taken by the British and put to trial by them. I have adventured my life in endeavoring to obtain the liberty of my countrymen, and am a willing sacrifice to their cause.*

Missouri Compromise also said that the rest of the Louisiana Purchase territory that was north of Missouri's southern border was to remain free. (Look at the map on page 151. This is complicated but important.) The Missouri Compromise kept some of the anger between North and South under control...for a while.

Meanwhile, in the first years of the 19th century, most European countries abolished—ended—slavery. Those countries began to criticize the United States for allowing it. Some people in the North—who for a long time hadn't seemed to care about it—started speaking out against slavery. By 1840 there were said to be about 2,000 abolitionist societies in the North. While some Northerners and Southerners talked of gradually freeing the slaves and even paying the slave owners, the abolitionists wanted to do away with slavery at once. They didn't think anyone should be paid for owning someone else. (As it turned out, it would have been much cheaper to pay the slave owners than to go to war.)

When the abolitionists published newspapers and books that attacked slavery, Southern postmasters wouldn't let them be delivered. (Southern whites were losing their liberty. They couldn't read what they wanted.)

Don't think this was a case of good Northerners and bad Southerners. Although slavery had been outlawed in the Northern states, many people in the North were treating blacks poorly too. Northern blacks were rarely given the rights of citizens: in most places they weren't allowed to vote or serve on juries. In the North, blacks often held the worst jobs in the community, and black children were usually not allowed in white schools.

Many white Northerners hated the abolitionists. Some Northern industries depended on Southern business. Because of that, some Northerners didn't want to upset the South. Other Northerners were just afraid of change. (And what the abolitionists were demanding was a major change in the United States.) Unfortunately, there are always people who fear new ideas. In the North, abolitionist presses were burned and destroyed and one abolitionist was actually murdered.

In the South, the abolitionists were really hated. When Northerners talked about abolishing slavery it made white Southerners furious. They were the ones whose lives would be changed. They didn't think it was the Yankees' business, and they said so. They didn't want to hear criticism of their beloved South. They didn't want to be told they were doing something immoral. Many white Southerners believed their liberty and property were being threatened by outsiders. They

**When Missouri wanted to become a state, James Tallmadge (left), a New York congressman, proposed a constitutional amendment to ban slavery there. The Southern states wouldn't agree to that, so Senator Jesse Thomas of Illinois (right) introduced an amendment to forbid slavery north of 36°30'—except in Missouri. Then Maine was made a nonslave state. That was the Missouri Compromise.**

**These were** the 12 free states in 1820 at the time of the Missouri Compromise: Connecticut, Illinois, Indiana, Maine, Massachusetts, New Hampshire, New Jersey, New York, Ohio, Pennsylvania, Rhode Island, and Vermont. And these were the 12 slave states: Alabama, Delaware, Georgia, Kentucky, Louisiana, Maryland, Mississippi, Missouri, North Carolina, South Carolina, Tennessee, and Virginia. According to the 1820 census, 5.1 million Americans lived in the free states and territories; 4.4 million lived in the Southern slave states.

*Would secession by New England have helped the slaves? Would it have prevented war?*

didn't worry about black liberty or property.

You can see this was an argument that was heating up. The abolitionists wrote and printed newspapers and books. Former slaves began to speak out and tell their stories. The abolitionists got angrier and angrier. Some abolitionists were so outraged by slavery they suggested that New England secede from the Union. That means they wanted to separate themselves from the other states. They wanted to form their own country. Some people in the South began saying the same thing. They wanted to secede and form their own country. These people were serious. No good would come of this.

## They Would Be Heard

*On January 1, 1831, a white Massachusetts man, William Lloyd Garrison, began publishing* **The Liberator.** *It soon became the leading abolitionist newspaper. Just one day before the first issue was printed, Garrison received a letter from James Forten (see p. 137 about Forten). This is what he said in his letter:*

I am extremely happy to hear that you are establishing a Paper in Boston. I hope your efforts may not be in vain; and that the "Liberator" be the means of exposing more and more the odious system of Slavery....Whilst...the spirit of Freedom is marching with rapid strides and causing tyrants to tremble, may America awake from the apathy in which she has long slumbered.

*In the first issue of* The Liberator, *Garrison wrote these famous words:*

I do not wish to think, or speak, or write with moderation. No! no! Tell a man whose house is on fire, to give a moderate alarm...but urge me not to use moderation in a cause like the present. I am in earnest—I will not equivocate—I will not excuse—I will not retreat a single inch—AND I WILL BE HEARD.

*Frederick Douglass, whom you will read about in the next chapter, was an eager reader. He had this to say about* The Liberator:

It became my meat and drink. My soul was set on fire. Its sympathy for my brethren in bonds— its scathing denunciations of slaveholders—its faithful exposures of slavery—its powerful attacks upon the upholders of the institution—sent a thrill of joy through my soul, such as I had never felt before!

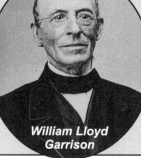

**William Lloyd Garrison**

# 31 Frederick Douglass

"You have seen how a man was made a slave," wrote Douglass. "You shall see how a slave was made a man."

What was it like to be a slave? Most white Americans—especially those in the North—didn't know. And then, in 1841, a tall, handsome man, a runaway slave whose name was Frederick Douglass, spoke up at an abolitionist meeting on Nantucket Island, near Boston. "I felt strongly moved to speak," Douglass wrote later. But he hesitated; his legs shook. "The truth was, I felt myself a slave, and the idea of speaking to white people weighed me down."

Yet he found the courage to speak out. When he did, eloquent words poured from his mouth. Frederick Douglass just told his own story: how he had lived and what he had seen. That was enough to send chills down the backs of his listeners.

Douglass soon became famous. He traveled from Nantucket Island to Indiana as a speaker for the Massachusetts Anti-Slavery Society. Northerners wanted to hear this man who spoke so well and told of his life as a slave. When he started an abolitionist newspaper, whites and blacks subscribed. Then he wrote a book and called it *Narrative of the Life of Frederick Douglass, an American Slave.* Here is some of his story:

> I never saw my mother...more than four or five times in my life....She made her journeys to see me in the night, travelling the whole distance on foot [12 miles], after the performance of her day's work. She was a field hand, and a whipping is the penalty of not being in the field at sunrise....I do not recollect of ever seeing my mother by the light of day. She was with me in the night. She would lie down with me, and get me to sleep, but long before I waked she was gone....She died when I was about seven years old....I was not allowed to be present during her illness, at her death, or burial.

**Give [a slave] a bad master and he aspires to a good master; give him a good master, and he wishes to become his own master. Such is human nature.**
—FREDERICK DOUGLASS

**A city slave is almost a freeman, compared with a slave on a plantation.**
—FREDERICK DOUGLASS

**People in general will say they like colored men as well as any other, but *in their proper place*. They assign us that place; they don't let us do it ourselves nor will they allow us a voice in the decision. They will not allow that we have a head to think, and a heart to feel and a soul to aspire....That's the way we are liked. You degrade us, and then ask why we are degraded—you shut our mouths and then ask why we don't speak—you close your colleges and seminaries against us, and then ask why we don't know more.**

—FREDERICK DOUGLASS

Young Frederick was sent to Baltimore to be a companion to a little white boy. For a slave, that was a lucky break. That chance, he said, "opened the gateway to all my subsequent prosperity." His new mistress was young and kindly. She had never had a slave. She began to teach the eager boy to read—until her husband saw her doing it and ordered her to stop. Reading, said the master, "would forever unfit him to be a slave."

The man was right. Through reading, a slave might learn about liberty and equality. The mistress stopped teaching Frederick. In fact, she did everything possible to keep books and magazines away from him. She began to turn mean. "Slavery proved as injurious to her as it did to me," Douglass wrote.

When Frederick's mistress was forbidden to teach him to read, he realized that the aim was to keep blacks ignorant. "From that moment," he wrote, "I understood the pathway from slavery to freedom."

But it was too late. Frederick had been bitten by the learning bug. Now he became determined to learn to read and to write too. He had to be clever to do it. He was often sent on errands where he met poor white boys. They were hungry for food; he was hungry for knowledge. They went to school and could read; he had extra food. He was soon trading bread for reading lessons. Sometimes he teased his white friends. "I can write better than you," he said, knowing he couldn't. After they proved they could write their letters, he had something he could copy.

Then he was sent away. The Baltimore people had just rented him. His owner took him back and sent him to a new master. This one was cruel, very cruel. Now he was beaten with a whip until he was bloody and scarred. He was not given enough to eat. He was sent into the fields to work long, long hours. He saw all the terrible things that happen when one person has complete power over another. "But for the hope of being free," Douglass wrote later, "I have no doubt but that I should have killed myself."

What happened to him next is all put down in his book—and it is exciting. But he didn't tell how he escaped to freedom. If he had, the slave catchers would have known how to capture others who might use the same route. He did tell the names of all the slave owners who had used and abused him. He told where they lived and all about them. That took great courage. He was still a runaway, and he knew slave catchers might come after him.

Frederick Douglass kept telling people this simple truth: *Justice to the Negro is safety to the nation.* (What do you think he meant by "safety to the nation"? Was he hinting at a slave uprising? Or did he think democracy might be at risk in a slave nation?) It was too bad more people didn't listen to him and ask themselves questions. The country would pay an awful price for its injustice and bigotry.

He also said, "You may rely upon me as one who will never desert the cause of the poor, no matter whether black or white." And he never did. He fought for human rights for all. In his later years he fought to get the vote for blacks and for women; he spoke out against the mistreatment of Chinese immigrants and American Indians; he worked for better schools for all.

Always he had the courage to stand up for his beliefs—well, not always; once it took all his strength to sit for those beliefs. He was in a railroad car and was asked to leave because he was black. He wouldn't budge. A group of white men tried to make him go. Douglass held on while they pulled the railroad seat out of the floor of the car. It was astounding how far hatred was taking some people.

Frederick Douglass became an adviser to President Abraham Lincoln, and a giant figure of American history. Do you think anyone knows, or cares, what happened to the bullies who threw him and the seat from the train?

**In 1835, Douglass traveled west to speak about abolition. In Pendleton, Indiana, he was stoned and beaten by a mob. His hand was broken and was never again as dexterous as it had been.**

### For Captive Millions

*Some Americans didn't think much about the Declaration of Independence or the ideals of the Founders, but slaves and free blacks understood just how precious freedom was. In 1837, James Forten, Jr., spoke at the Ladies' Anti-Slavery Society:*

My friends, do you ask why I thus speak? It is because I love America; it is my native land; because I feel as one should feel who sees destruction, like a corroding cancer, eating into the very heart of his country, and would make one struggle to save her;—because I love the stars and stripes, emblems of our National Flag—and long to see the day when not a slave shall be found resting under its shadow; when it shall play with the winds pure and unstained by the blood of "captive millions."

# Walking Across the Map

If you don't know where you are, you'll never figure out where you are going. Does that make sense? Well, even if it doesn't, believe me, if you want to understand history you need to know some geography. And that means you need to do some map reading.

Let's look at the land that is now the United States and see what it was like in Andrew Jackson's time. (When was that?) Notice the original 13 states along the East Coast. Much of that land (except in New England) is flat, tidewater land. Going west, you come to piedmont —rolling foothills. Then we have mountains—the ancient and beautiful Appalachian chain. Go over the mountains: to the north is the Northwest Territory. New Englanders are settling the northwest and bringing their tidy, hardworking habits with them. Indiana and Illinois are now states. Slavery is not allowed in the Northwest Territory.

Look south. Hundreds of thousands of people have poured into the lands of Kentucky and Tennessee in the years since 1776. Kentucky became the 15th state and Tennessee the 16th. People are getting to those new states by going west on the Wilderness Trail that Daniel Boone helped cut through the Cumberland Gap.

Boone isn't there to greet them. When white settlers began filling the lands watered by the Ohio River and its tributaries, Boone moved west— across the Mississippi to Missouri. That was where he died, in 1820, the year before Missouri became a state.

Other over-the-Appalachian lands to the north and south have settlers, but not enough yet to claim statehood. They will become states when—according to the rules set down in the Northwest Ordinance— they have enough people. The southern territories will become slave states.

When the over-the-mountain people have crops or furs they want to sell, how do you think they get them to market? What is the easiest method of transportation in the early 1800s? Hint: America is a land of rivers.

Yes, boat transportation. People try to settle as near to a river as possible. They climb onto riverboats and float away. Two huge rivers do a lot of the transporting. They are the Ohio and Mississippi rivers, and they are like superhighways. Take a look at the map. Find Pittsburgh. Do you see the Ohio River? Put yourself on a flatboat. You can bring your possessions. A houselike structure stands on one end of the boat. Some of the cargo will go in there for protection. When it rains, you can squeeze inside, too. But you'll spend most nights sleeping under the stars, or under a tent at the river's edge. If there is a danger of pirates, you may sleep by day and float quietly at night.

Now let's head downriver. Keep checking that map—this river does a lot of wending and winding. Do you see Cincinnati and Cairo? At Cairo we'll float into the great Mississippi, then down to New Orleans and on to the Gulf of Mexico. We'll continue by ship around Florida and up the East Coast. It is a long way around, but faster and easier than overland by cart. Which tells you what the roads are like.

Check the map again. Put a finger on Cincinnati. It is cheaper to send a barrel of goods from Cincinnati to New York by water than it is by land.

Now look at the Louisiana Territory, that land President Jefferson bought in 1803. You can see it clearly on the map. It wasn't clear at all in 1803, and it still isn't clear to most people in 1829; no one knows its exact size.

Now keep your eyes on the map and head north and west. Notice the Oregon Country. The United States, England, Russia, and Spain are all claiming Oregon. Oregon is rich in furs, and everyone wants otter and beaver skins.

Now look south, down the West Coast. A powerful nation has planted roots in California. It is Russia. Russian traders settled in the Aleutian (uh-LOO-shun) Islands (the bony finger of islands that stretches off from Alaska) about the time that George Washington was settling in at Valley Forge. Then the Russians went down to a spot just north of today's San Francisco and built a fort: Fort Ross. (Actually, it was named *Rossiya*, which is how the Russians say "Russia." The Americans soon shortened it to *Ross*.)

Fort Ross is a fur-gathering spot. Some Russians have made a fortune selling otter skins in China and Europe. They are hunting those seagoing animals almost to extinction. Spain doesn't like the idea of the Russians being in this area.

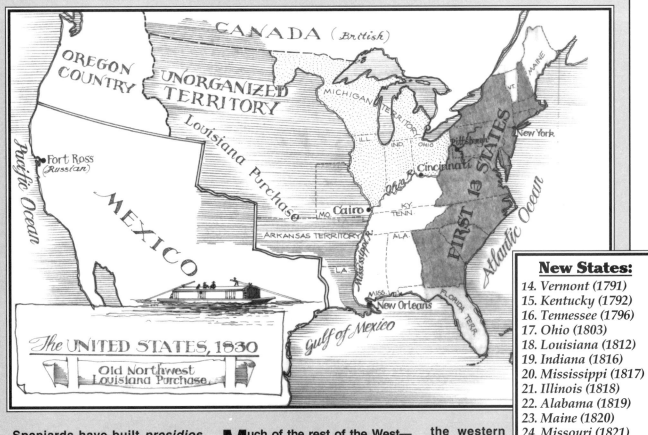

The UNITED STATES, 1830

Old Northwest
Louisiana Purchase

**New States:**

14. *Vermont (1791)*
15. *Kentucky (1792)*
16. *Tennessee (1796)*
17. *Ohio (1803)*
18. *Louisiana (1812)*
19. *Indiana (1816)*
20. *Mississippi (1817)*
21. *Illinois (1818)*
22. *Alabama (1819)*
23. *Maine (1820)*
24. *Missouri (1821)*

Spaniards have built *presidios* (forts) to prevent the Russians from expanding in California. The Americans wish that both Russia and Spain would keep to their home continents.

Events in Russia will make the Russians leave. Before they do—in 1842—they will turn Fort Ross into a charming place, with fine houses, a chapel, storehouses, vineyards, and even a greenhouse of glass. A French visitor says the place has a "wholly European air." Russians, Aleuts, and Indians live there with their families—about 400 people in all.

Much of the rest of the West—from California to Colorado and south to what is Texas today—is now part of Spanish Mexico. (It was called New Spain, and was controlled by Spain, until 1821, when Mexico won its freedom from Spain. You can read about what happened to that land in Book 5 of *A History of US*.) Spanish-speaking priests and ranchers have settled here. The priests are teaching the Indians about Christianity. The Indians are doing most of the hard work of cooking, planting, building, and raising animals in the Mexican settlements. No one wishes it, but

the western Indians are suffering terribly from European and African diseases. Much of the native population is being wiped out.

In the United States, people are eyeing these western regions. President Monroe announces his Monroe Doctrine (in 1823) to warn Spain and Russia to stay away from this continent. Some Americans would like their nation to include all of Mexico and Canada. James Monroe, John Quincy Adams, and others expect that someday the United States will stretch from coast to coast.

151

# 32 Naming Presidents

In 1790, there were still only 13 United States, but the Northwest Territory was American and being settled fast. In the next 10 years, Vermont, Kentucky, and Tennessee became states (Florida still belonged to Spain). The big jump in territory comes in 1803, when the Louisiana Purchase almost doubled the country's size. Ohio became a state that year. In 1812 Louisiana followed, and then, in quick succession, Indiana, Mississippi, Illinois, and Alabama.

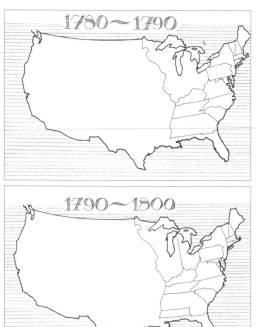

**Jackson's enemies said that the president trampled on the Constitution and did whatever he wanted. They called him King Andrew I.**

Do you remember who was president the last time I was on that subject?

Are you good at memorizing the names of presidents? I usually get stuck somewhere in the middle. But, as with the states and their capitals, it is nice to know them.

Now, I'm sure you can name the first seven presidents. You know them all. Yes, you do.

Of course you know number one: the man who was said to be "first in war, first in peace, and first in the hearts of his countrymen."

The second president had a wife named Abigail. They were New Englanders, from Massachusetts. (Presidents one, three, four, and five were all from Virginia.) Clue: the second president was an important man at the Continental Congress in 1776. He helped persuade the best writer there to write a declaration telling the British to go home.

The man who wrote that famous declaration was president number three. He was the most versatile of the Founding Fathers. That means he could do many different things, and do them all well. He wrote the

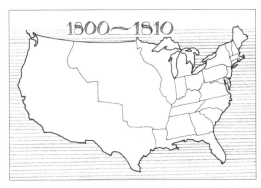

1800~1810

Virginia Statute for Religious Freedom and founded the University of Virginia.

President four was a good friend of president three. He was called the "Father of the Constitution." He wrote the Bill of Rights. His wife's name was Dolley.

President number five was the last Virginia president to wear knee pants and buckles on his shoes. An important policy—called a *doctrine*—bears his name. That doctrine told Europe's nations to keep their hands off the two American continents.

President number six was the son of president two. They were the only father and son to be presidents. Number six was serious, dedicated, and very smart.

1810~1820

Number seven was a general who won some big battles. He was a man of the people, but he didn't seem to think that Native Americans or slaves were Americans.

If you can't name the first seven presidents I'm not going to name them for you—not here, anyway. (But if you really don't remember them, you will find their names in earlier chapters of this book.)

Those first seven were outstanding. Now come the next eight, who were not. (Most people think that the president after that—number 16—was the best of all.) I'm going to list those next eight presidents and just tell you a few things about each of them. Some of the things you will read are serious and some are silly, but all are true. I would like to tell you more—I really should tell you more—but there is so much to write about in history that choices have to be made. If you want to know more about the presidents to come, you

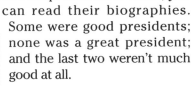

1820~1830

can read their biographies. Some were good presidents; none was a great president; and the last two weren't much good at all.

### President number 8
*Martin Van Buren (1837–1841)*
    Van Buren was the first president who was born a citizen of the United States. The

1830~1840

## Presidential Facts
### (Some are funny!)

☞ **Van Buren** didn't get reelected in 1840. Eight years later he tried again as head of a new party, the Free Soil Party. It had an important cause: to keep slavery out of the western territories. In Virginia the Free Soilers got only nine votes. They said there was fraud (cheating). One Virginian agreed. He said, "Yes, fraud! And we're still looking for that son-of-a-gun who voted nine times!"

☞ **In 1840,** Harrison supporters pushed a huge paper ball covered with campaign slogans from city to city. They added a phrase to the language: "Keep the ball rolling."

1820 was a big year. Then, under the Missouri Compromise, Maine became the 23rd state (no slavery allowed); in 1821 Missouri became the 24th—and the 12th slave state. And that same year we bought Florida from Spain, too.

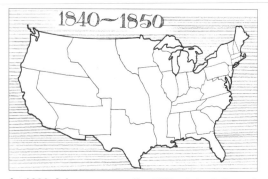

In 1836 Arkansas was the 25th state, followed next year by Michigan. Those old Spanish lands Florida and Texas became states in 1845. Then came Iowa and Wisconsin in 1846 and 1848.

other presidents weren't born Americans? No, they weren't. They were all subjects of the British king when they were born.

Martin Van Buren was a hardworking president who believed political parties were necessary and important. He helped turn the old Democratic-Republicans into the modern Democratic Party. He was a good friend of Andrew Jackson, but he was never as popular as the general.

Van Buren was born in Kinderhook, New York. He was called Old Kinderhook, or O.K., by his supporters. The term *okay* (no one is quite certain where it came from) became popular in the election of 1840; it was used by Van Buren's supporters to describe him, and then by others to describe anyone who is—okay!

**President number 9**
*William Henry Harrison (1841)*
Harrison, who defeated the Prophet at Tippecanoe, was an aristocrat and the son of a Virginia governor. William Henry went to college, studied medicine, and became an army officer.

But aristocrats were out of favor in 19th-century America. So Harrison's campaign managers made him sound like an old log-cabin boy who came up the hard way. (He hadn't.) They made his opponent—Van Buren—sound like the aristocrat. (He wasn't.) They even accused Van Buren of being so fancy he put a bathtub in the White House! (He didn't.)

Well, it worked. Harrison was elected president. He rode a white horse to his inauguration, but refused to put on a coat or hat—even though the weather was bitterly cold.

Then Old Tippecanoe (that was his nickname), gave the longest inaugural address in history. He spoke for nearly two hours and caught a cold. It developed into pneumonia, and that was the end of William Henry Harrison. He was the first president to die in office. He was chief executive for only 31 days.

### President number 10
### *John Tyler (1841–1845)*

When William Henry Harrison ran for the presidency, his election campaign slogan was "Tippecanoe and Tyler too." John Tyler, the vice-presidential candidate, was *Tyler too*. He was the first vice president to take office because of the death of a president. Tyler, another Virginian, played the violin, loved to dance, spoke softly and had good manners. He had more children than any other president: 14. He was said to have been playing marbles with one of his sons when a messenger came with word that Harrison had died and he was president. Tyler and Harrison were elected as members of the new Whig party. (The Whigs were those who didn't like Andrew Jackson. Many of them were former Federalists. Henry Clay and Daniel Webster were two Whig leaders.) But Tyler fought with the leaders of the party, and after he had been in office five months they said he was no longer a Whig.

Florida was named a state while Tyler was president.

### President number 11
### *James Knox Polk (1845–1849)*

President Polk, who was from Tennessee, didn't approve of dancing and didn't like music, except for hymns. He thought having fun meant wasting time, and he didn't like to waste time. No refreshments were served at White House receptions when he was president. He said, "I am the hardest working man in the country." His hard work paid off: the Oregon and California territories became part of the United States while Polk was president, and Iowa, Texas, and

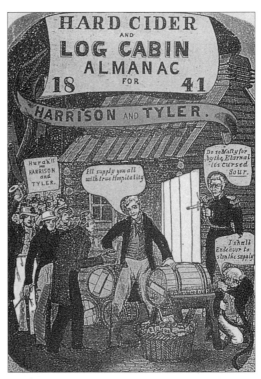

In this poster from Harrison and Tyler's election campaign, Martin Van Buren tries vainly to stop the happy flow of Harrison's cider while Jackson encourages him—"Do so Marty for by the Eternal it's cursed sour," he says.

## Presidential Facts
### (Continued)

☞ **During the** first 52 years of United States history there were eight presidents. During the next 20 years there were eight presidents.

☞ **In 1841** the United States had three presidents. Martin Van Buren's term ended on March 4; on that day William Henry Harrison was inaugurated as the new president. A month later, Harrison was dead and John Tyler was president number 10.

Wisconsin became states. He fought a war with Mexico. Polk kept a diary (which you can read). He wouldn't run for a second term.

### President number 12
### *Zachary Taylor (1849–1850)*

Taylor was a brave, patriotic general who was called "Old Rough and Ready" by his troops. He was born in Virginia, but moved to Louisiana. He owned slaves, but he didn't want slavery to spread into the western territories. When Southerners talked of secession (leaving the Union), he said he would send troops south and stop them.

Taylor had no political experience before he became president. He had never even voted in a presidential election. When the Whig party leaders nominated him, they sent a letter collect to tell him the news. The receiver pays for a collect letter. Taylor wouldn't pay. It was a few weeks later that he learned of his selection.

Taylor was an intelligent man who knew how to make decisions. He might have been a good president, but he died in office after serving only 16 months.

### President number 13
### *Millard Fillmore (1850–1853)*

Millard Fillmore was a New Yorker and the second vice president to take over after a president died. When he was a boy, he was indentured to a clothmaker. Fillmore installed the first kitchen stove in the White House. The cook couldn't figure out how to work the stove. President Fillmore went to the U.S. Patent Office, read the patent for the kitchen stove, and came back and taught the cook how to use it.

### President number 14
*Franklin Pierce (1853–1857)*

Pierce was a New Hampshire man who graduated from Bowdoin College in Maine. It took the Democratic party 49 rounds of voting to choose him at their presidential convention. He was a "dark horse" candidate. All the likely candidates got defeated, until finally Pierce—who wasn't well known—was left. Although he was a Northerner, he did not object to slavery, and he disliked the abolitionists.

Two months before Franklin Pierce was inaugurated, his 11-year-old son, Bennie, was killed in a train accident. The Pierces were sad all the rest of their lives. Pierce was not a strong president.

### President number 15
*James Buchanan (1857–1861)*

James Buchanan was the only bachelor president. His niece stayed at the White House and hosted his parties. His inauguration was the first to be photographed. Buchanan came from Pennsylvania.

In his time the arguments about slavery grew fierce. He didn't do anything to try to make things better. In fact, he didn't do much of anything as president. When Congress passed a bill that would have created some colleges, Buchanan vetoed the bill. He said that the country didn't need more education. There were already too many educated people, said Buchanan.

President Harry S. Truman (who lived in another century, but knew his history) called him "an old fool."

Things were a mess when James Buchanan stepped down from the presidency. That made the job difficult for the next man in the office: Abraham Lincoln.

## Presidential Facts
### (Concluded)

☞ **When Franklin** Pierce was running for president, his Whig opponents made fun of his military record by distributing a miniature book—an inch high by half an inch wide—entitled *The Military Services of General Pierce.*

☞ **Buchanan was** Polk's secretary of state. Former president Andrew Jackson protested. "But, General," said Polk, "you yourself appointed him minister to Russia in your first term." "Yes, I did," said Jackson. "It was as far as I could send him out of my sight, and where he could do the least harm. I would have sent him to the North Pole if we had kept a minister there!"

California became a state in 1850, Minnesota in 1858, Oregon in 1859, and Kansas in 1861. That made 34 states. So far, there weren't enough people between Kansas and California to turn that land into states. (It will take 98 more years before we get to number 50. What state is that?)

The first seven presidents are easy, but how are you going to remember presidents 8 through 15? These are your choices:

1. Forget them (worst choice).
2. Say them over and over (okay choice).
3. Find a memory system and remember them easily (best choice).

What is a memory system? Anything that will help you remember. Memory systems are called *mnemonics*. Don't pronounce the *m* and you can say it: nuh-MON-ix.

Here is a mnemonic idea for remembering the presidents. Take the first initial of each of their last names—VHTPTFPB—and make a sentence with them. Any sentence will do. Here is one as an example:

**Very Heavy Tennis Players Throw Fast Purple Balls.**

Does that make sense? Of course not. What it means is: Van Buren, Harrison, Tyler, Polk, Taylor, Fillmore, Pierce, Buchanan.

## Yankee Doodle Presidents *(sung to the tune of "Yankee Doodle")*

*George Washington, first president, by Adams was succeeded.
Thomas Jefferson was next, the people's cause he pleaded.
James Madison, he then came forth, to give John Bull a peeling.
James Monroe was next to go in the "era of good feeling."*

*'Twas John Q. Adams then came in and, after, Andrew Jackson,
He licked the British at New Orleans to our great satisfaction,
Van Buren then was next to chair, then Harrison and Tyler—
The latter made the Whigs so mad they almost burst their b'iler.*

*We next elected James K. Polk, the worst that then did vex us
Was, should we fight with Mexico and take in Lone Star Texas?
Then Taylor was our leader, but soon had to forsake it,
For Millard Fillmore filled it more, Frank Pierce then said, "I'll take it."*

*At problems James Buchanan yawned and Lincoln then was picked,
He found the troubles of the day were anything but licked.
Old Johnson had a rough, rough time, the Senate would impeach him,
But as it took a two-thirds vote, they lacked one vote to reach him.*

**T**hese are the silhouettes of four of the first 16 presidents. See if you can figure out which they are. (The answers are upside down on the right if you give up.)

*John Bull* is another name for England, as *Uncle Sam* is for the U.S. The *Whigs* were the political party that inherited Federalist concerns. *B'iler* is "boiler" with a Scots-Irish accent.

The presidents are, clockwise from top left: Martin Van Buren, William Henry Harrison, John Quincy Adams, John Tyler.

# 33 A Triumvirate Is Three People

Daniel Webster said, "We may be tossed upon an ocean where we can see no land.... But there is a chart and compass for us to study....That chart is the Constitution."

*Triumvirate* comes from the Latin words *trium* (which means "of three") and *vir* (man). There were times in ancient Rome when government control was divided among three men so that no one person would be too powerful.

If you could pick a time in all of history to sit in the galleries of Congress and listen, you would do well to consider the first half of the 19th century. It was an age of oratory. The House and Senate echoed with eloquence. But among the silver-tongued speechmakers, three were outstanding. Each of those three wanted to be president, each tried to be president, each was more brilliant than any of the eight men who won the office after Andrew Jackson. Even Patrick Henry might have met his match in a debate with any of those three. Their names were Henry Clay, John C. Calhoun, and Daniel Webster.

They say that when Daniel Webster spoke even his enemies came to marvel. In New England, where he was born, they called him the "godlike Daniel." He was a heroic figure of a man, sturdy, with dark hair and flashing eyes. "No man can be as great as that man looks," said someone who saw him. But it was his deep, strong voice that most people remembered. Ralph Waldo Emerson, a writer who knew him, said that compared to others, Webster was like "a schoolmaster among his boys." He would have made a powerful actor; he was a powerful senator.

Being a Yankee, Daniel Webster supported the interests of the Massachusetts factory owners, bankers, and shipowners. He hated slavery and used strong words to attack it. He loved the Union and did everything he could to protect it. He wanted to be president.

So did Henry Clay. Henry Clay, ah, Henry Clay. There was a charming man. He was called "the great compromiser." Some say that if Clay

**Daniel Webster** longs to be president; the ambition consumes him. Late in his life, when he realizes he will never make it, he is angry and bitter. This is what he says: "I have given my life to law and politics. Law is uncertain and politics is utterly vain."

**A big** argument in Congress was over states' rights versus federal power. Which should be stronger? Some people are still arguing about that.

**Andrew Jackson** was head of the Democratic Party. (It had evolved from Thomas Jefferson's old Democratic-Republican Party.) But the people who couldn't stand Jackson (and there were quite a few of them) needed a party to express their ideas. Henry Clay was their leader. He founded the Whig Party. It was named for the English political party that had opposed the king. Clay accused Jackson of acting like a king.

**Have you** ever heard someone called a "bigwig"? The expression really should be *Big Whig*. It was used to describe members of Henry Clay's Whig Party. Most of the wealthy and influential people of the time were Whigs—just the kind of people that today we call bigwigs.

had been president there would have been no Civil War. Perhaps. He certainly tried to get the job. He ran for president five times and was well qualified for the position. He was a senator when he was 29 (even though the Constitution says you must be 30); after that he became speaker of the House of Representatives; secretary of state under John Quincy Adams; and a senator again.

Clay was born in Virginia. At the College of William and Mary he studied with Jefferson's great law professor, George Wythe, and he heard Patrick Henry debate. Then he moved to Kentucky, which was frontier country. In Congress he was known as "Harry of the West."

What a speaker he was! He could talk right out of his head, without anything written down, and make good sense and never bore a listener. His speeches were said to be "magnetic." No doubt about it, he attracted people and held them in a web of mellow words and strong ideas.

Tall, with a fine-featured long face, blue eyes, and a flashing smile, Henry Clay often wore a flowered vest, a high white collar, and a white tie. He liked having a good time and was friendly; there were many who adored him. It was his intelligence that scared his enemies. They knew they could not control him as they could some of the weaker men who became president. Clay spoke out against slavery, and yet he had slaves himself. Like Daniel Webster, Clay loved the Union and would do everything he could to save it.

John Calhoun was another man who wanted to be president. Calhoun, from South Carolina, had gone to college and law school in New England. Still, he was no friend of Daniel Webster's friends. Calhoun, too, was dark-

eyed, handsome, and powerful of mind and tongue. But he had none of the humor or hearty spirit of Webster. Calhoun was serious, and he had a serious mission. He needed to explain slavery and make people believe, as he did, that slavery was necessary and even good. He thought what he was doing was right. Calhoun loved the Union, but he loved the Southern way of life even more. And he was afraid that way of life was in danger. Once,

**Henry Clay said, "The Constitution of the United States was made not merely for the generation that then existed, but for posterity—unlimited, undefined, endless, perpetual posterity."**

the South had been more powerful than the North. By 1820, that was changing. If the western territories joined the Union as free states—if they voted with the North—then they might vote to end slavery. It would be the end of the South that Calhoun knew and loved. Calhoun wanted the new states to become slave states. He worried about minority rights. But the minority he had in mind was the South and its white citizens. It never seemed to occur to him that blacks, too, were a minority.

Calhoun fought the tariffs the North wanted. A tariff is a tax or duty on goods brought from a foreign nation. If foreign goods are taxed, they become more expensive than untaxed goods made at home. High tariffs helped the Northern industries that were just beginning to grow. Tariffs hurt the Southerners who sold cotton in Europe and bought manufactured goods there. Calhoun said if a state believed a law was unconstitutional, it didn't have to obey that law. He said the Southern states should not collect tariffs. John Calhoun was Andrew Jackson's vice president, but Jackson was furious when he heard Calhoun's ideas on not collecting tariffs, and he didn't agree with Calhoun's hard stand on slavery.

There were powerful differences in America, and the Senate was just the place to argue them. The small Senate chamber had a high, domed ceiling that reflected even a hushed voice. The 48 senators sat at desks in semicircular rows, each row higher than the one in front of it. The senators had no clerks, or assistants, or offices (as they do today), so all their work was done at their Senate desks. Because of that, the Senate was often a noisy place. But when one of the great orators spoke, the room became still and the balcony quickly filled with spectators.

"There never has yet existed a wealthy and civilized society in which one portion of the community did not...live on the labor of the other," said John Calhoun.

## Daniel, You're Wrong!

*In 1829, Daniel Webster opposes the establishment of a cross-country mail route. This is what he says in Congress of the western part of the country:*

**W**hat do we want with this vast, worthless area? This region of savages and wild beasts, of deserts, of shifting sands and whirlwinds of dust, of cactus and prairie dogs? To what use could we ever hope to put these great deserts, or those endless mountain ranges, impenetrable and covered to their very base with eternal snow? What can we ever hope to do with the western coast, a coast of three thousand miles, rockbound, cheerless, uninviting, and not a harbor on it?

**Despite their** differing ideas, Webster, Clay, and Calhoun respect each other. Calhoun says of Henry Clay: "He is a bad man, an imposter, a creator of wicked schemes. I wouldn't speak to him, but, by God, I love him."

You'll read more about Sam Houston (above) and John Quincy Adams (below) in Book 5 of *A History of US*.

Do you think you might have liked being there just to listen? Besides Clay, Calhoun, and Webster, there was the senator from Texas, Sam Houston, who spoke boldly for what he thought was right—and, though he lived with slavery, he didn't think it right. He didn't agree with John Calhoun and said so. Texas was a slave state, and what Houston said was not popular in Texas. He lost his seat in the Senate, as he knew he would.

And there was Senator William Seward of New York, the man who bought Alaska for the nation. (That would happen in 1867.) Seward was an uncompromising enemy of slavery. The Constitution allowed slavery. Seward said, "There is a higher law." What did he mean?

But the Senate wasn't the only place where you could hear eloquent speakers. John Quincy Adams was in the House of Representatives. That's right, JQA was in Congress—elected to the House after Andrew Jackson beat him in his second try for the presidency.

The people of Plymouth, Massachusetts, asked him to serve, and Adams said he would if he could vote his conscience and not worry about what they thought. That was a typical John Quincy Adams request; the people of Plymouth weren't surprised a bit, and were proud to send him to Washington.

"Do you find it hard serving in Congress after being president?" he was asked.

He didn't find it hard at all. Liked it better than anything he had done before. For John Quincy, serving the public was reward enough. "Old Man Eloquent" was what they called him in Congress, and, because he always made sense, people stopped whatever they were doing to listen when he spoke. In Congress he was a leader in the fight against slavery. And it was in Congress, in 1848, that he had a stroke and died—which may have been the way he wanted to go.

But it was a debate in the Senate, in 1830, that I want you to hear. Some people say it was the greatest debate ever heard in the Senate. Some people say it was the greatest in United States history. Now you can decide for yourself.

**"Slavery is the great and foul stain upon the North American Union," said John Quincy Adams. He died doing what he thought was right.**

# 34 The Great Debate

"I never use the word 'Nation' in speaking of the United States," said John Calhoun. "We are not a Nation, but a Union, a confederacy of equal and sovereign States."

Let's take seats in the Senate gallery. From here we can look over the railing and see what is going on below. It is 1830, and Vice President John Calhoun is presiding. (The vice president leads the Senate but cannot give speeches or vote—unless there is a tie.) South Carolina's popular 38-year-old senator, Robert Young Hayne, is speaking. Hayne is slim, good-looking, and well respected. But some say that it is Calhoun who is really talking today. Because the vice president cannot speak in the Senate, they say that Hayne is just mouthing Calhoun's words. That isn't fair. Hayne is a also a brilliant man, and a fine speaker. Yet there is no question that Hayne looks up to Calhoun. Calhoun will nod and smile at him as he speaks.

Hayne is urging the West to unite with the South and oppose the North and its tariff. Together the South and West can dominate the nation, he says. Senator Hayne talks of "states' rights." It is that old argument of Patrick Henry's. Hayne believes the country was formed by and for the states—not the people in general. He fears a strong federal government; he wants each state to keep final power for itself. His arguments are clear and brilliant. Can anyone answer him?

Daniel Webster rises. His is a speech no one will forget. He will talk for two days, trying to defend the Union against the Southern congressmen

At the very time the great debaters are speaking in Congress, people at Fort Dearborn are making plans to turn their federal post into a city. It will be called Chicago. This same year (1830), in Philadelphia, Louis A. Godey begins publishing *Godey's Lady's Book*. It is the first successful women's magazine.

After leaving Congress, Robert Young Hayne becomes governor of South Carolina and then president of the Louisville, Cincinnati & Charleston Railroad Co. He sinks a lot of money into the railroad and loses it all. (Sure things aren't always so sure.)

**In 1832,** Thomas Jefferson Randolph, the grandson of the third president, presents Jefferson's emancipation plan in the Virginia assembly. It loses in a close vote.

**"The inherent right in the people to reform their government I do not deny," says Daniel Webster. "And they have another right, and that is to resist unconstitutional laws without overturning the government."**

who now say that their liberty is more important than the Union. They are beginning to talk openly of secession. That means leaving the Union and forming a separate new nation in the South.

Picture a big man dressed in a blue coat with brass buttons, a cream-colored vest, and a white tie. He uses his voice like a musician playing an organ. Sometimes the voice thunders and fills the hushed hall; sometimes the voice is soft and sweet and people need to concentrate so they won't miss a word. There isn't an empty seat in the Senate chamber. So many congressmen have come to listen that the House of Representatives cannot conduct its normal business.

What is this government of ours, Webster asks? Does it belong to the state legislatures, or to the people? And he answers his own question:

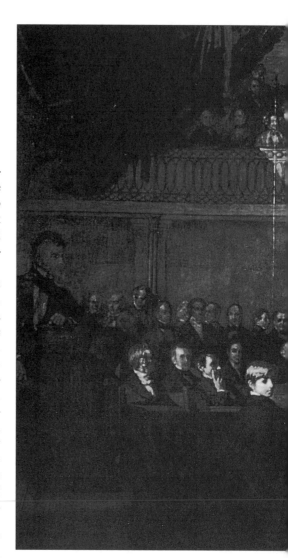

> *The Constitution is not the creature of the State government. It is, sir, the people's Constitution, the people's government, made for the people, made by the people, and answerable to the people. The people of the United States have declared that this Constitution shall be supreme law.*

The states have powers, he tells his hushed listeners, and he describes the powers the Constitution gives the states. He tells of the

checks and balances and limits on power. He is giving his audience a lecture on constitutional law. The Constitution is the great law of the land, he says. Who shall decide if a law is constitutional? For Webster there is only one answer: it cannot be the states; it must be the Supreme Court.

But it is secession that really worries him. The great orator uses patriotic phrases to fight that idea. Webster's speech will become so famous that children in school will be required to memorize long parts of it. *When my eyes shall be turned to behold for the last time the sun in heaven, may I not see him shining on the broken and dishonored fragments of a*

## My Country 'Tis of Thee

**S**amuel F. Smith is asked to write a hymn for a children's choir. He writes out words about our country and fits them to a tune in a German songbook. It is 1832, and Smith calls his hymn "My Country 'Tis of Thee." Later, he discovers he has chosen a melody that is also sung with the words "God Save the King"—which is the British national anthem.

*once glorious Union*, he pleads. He warns of civil feuds and the shedding of brothers' blood. How can his listeners know that he is predicting the future? Then he speaks of an idea of John Calhoun's—*Liberty first and Union afterwards*—and calls it folly. Finally, he ends with these ringing words: *Liberty and Union, now and forever, one and inseparable!*

The western senators are won over. Daniel Webster has crushed the Southern hope that the West and the South can band together. But Webster has not crushed John Calhoun.

Calhoun understands that if the South is to keep slavery, all the nation must approve of it. He must make the nation approve, or help form a separate Southern nation. Calhoun will develop a philosophy that makes slavery sound noble. Earlier Southern leaders, men like Jefferson and Madison, talked of slavery as an evil they were unable to control. Calhoun will call it a "positive good." Other Southern leaders will say it is God-inspired. Many people will believe them.

President Andrew Jackson is not one of them. He is a strange mixture of fighter, democrat, and practical politician. He is a slave owner, but he never suggests that slavery is good. Yet he doesn't say that

slavery is bad, either. He doesn't provide much leadership on this issue that is so important to the United States and its peoples.

But Jackson does put the nation ahead of his home region. When South Carolina refuses to collect the tariff, Jackson threatens to send an army to his native state. "I will hang John Calhoun," he fumes. He doesn't really mean that, but he and his vice president are now bitter enemies.

Soon after his debate with Daniel Webster, Robert Hayne gives up his seat in the Senate to become governor of South Carolina. As soon as that happens, John Calhoun resigns as vice president. If he can't be president, Calhoun would rather be a senator. He knows the nation's future is being decided on the floor of Congress. Calhoun is appointed to replace Hayne. For the next 20 years John Calhoun will speak for himself in the Senate. He will have much to say.

**"When men of high standing attempt to trample upon the rights of the weak, they are the fittest objects for example and punishment," said Andrew Jackson.**

# 35 Liberty for All?

**While Webster and Hayne debate in the Senate, a 21-year-old (who admires Henry Clay) moves with his family to Illinois. His name is Abraham Lincoln.**

How do you think the nation is doing? Are you discouraged? After all, there is slavery, some people are being horrible to the Native Americans, and there is much lawlessness in the land.

Does it look hopeless for freedom, democracy, and fairness?

Well, don't be discouraged. Actually, the United States is doing amazingly well. Those Founders—Franklin, Washington, Jefferson, Madison, and the others—never suggested that good, fair government would be easy to achieve. They set a splendid goal: a land where all were to have equal rights. They wrote an extraordinary plan of government: a constitution better than any known before.

Now it is the nation as a whole that is being tested. Nowhere before has a large, diverse nation tried to offer all its citizens freedom and equality. Doing that isn't easy. There are problems, big problems—but Americans will keep working toward that fine goal of justice for all. It will take time. It will take courageous people, but we will make great progress toward that goal. The slaves will be freed. Women will get the vote. Native Americans will gain the rights of U.S. citizens. Segregation will become illegal. Keep reading. You will see.

**Andrew** Jackson's last words before his death in 1837 are: "I expect to see you all in heaven, both white and black."

167

# Chronology of Events

**1789:** George Washington is elected first president of the new United States

**1790:** Potomac River site picked for the new capital

**1790:** death of Benjamin Franklin

**1790:** the first United States census

**1791:** Vermont becomes the 14th state

**1791:** Samuel Slater's spinning mill starts up in Pawtucket, Rhode Island

**1791:** Robert Carter III frees his slaves

**1792:** Kentucky becomes the 15th state

**1793:** Eli Whitney invents the cotton gin

**1796:** John Adams is elected second president

**1796:** Tennessee becomes the 16th state

**1798:** Congress passes the Alien and Sedition acts

**1799:** death of George Washington

**1800:** Thomas Jefferson is elected third president

**1801:** John Marshall appointed Supreme Court chief justice

**1803:** Ohio becomes the 17th state

**1803:** the Louisiana Purchase

**1803:** the Supreme Court case of *Marbury* v. *Madison* establishes the process of judicial review

**1804:** Alexander Hamilton killed in a duel with Aaron Burr

**1804–1806:** Lewis and Clark expedition

**1807:** Robert Fulton's steamship *Clermont* travels up the Hudson River from New York to Albany

**1808:** James Madison elected fourth president

**1808:** the slave trade to America becomes illegal

**1811:** William Henry Harrison defeats the Shawnee, led by the Prophet, at Tippecanoe

**1812:** Louisiana becomes the 18th state

**1812–1815:** in the War of 1812 the United States eventually defeats Britain once and for all

**1813:** Tecumseh is killed in battle

**1814:** Andrew Jackson defeats Creek Indians at Horseshoe Bend

**1814:** Francis Scott Key writes "The Star-Spangled Banner" to celebrate the British defeat at Baltimore

**1816:** James Monroe elected fifth president

**1816:** Indiana becomes the 19th state

**1817:** Mississippi becomes the 20th state

**1818:** Illinois becomes the 21st state

**1819:** Alabama becomes the 22nd state

**1820:** the Missouri Compromise

**1820:** Maine becomes the 23rd state

**1821:** Florida bought from Spain for $15 million

**1821:** Missouri becomes the 24th state

**1821:** Sequoyah devises a Cherokee alphabet

**1823:** President Monroe states the Monroe Doctrine

**1824:** the Erie Canal is completed

**1824:** John Quincy Adams elected sixth president

**1826:** Thomas Jefferson and John Adams both die on July 4

**1828:** Andrew Jackson elected seventh president

**1830:** the Indian Removal Act clears the way for sending Native Americans west of the Mississippi

**1830:** Daniel Webster replies to Robert Hayne in a Senate speech on the importance of the Union

**1830:** Peter Cooper builds a steam-powered train, the *Tom Thumb*, for the Baltimore & Ohio railroad

**1831:** William Lloyd Garrison begins publishing the abolitionist newspaper *The Liberator*

**1836:** Martin Van Buren elected eighth president

**1836:** Arkansas becomes the 25th state

**1837:** Michigan becomes the 26th state

**1838:** Cherokees walk the Trail of Tears westward

**1838:** Osceola dies in prison

**1838:** the slave Frederick Douglass flees the South

**1840:** William Henry Harrison elected ninth president

**1841:** John Tyler becomes 10th president when Harrison dies

**1844:** James K. Polk elected 11th president

**1845:** Florida becomes the 27th state

**1845:** Texas becomes the 28th state

**1846:** Iowa becomes the 29th state

**1846–1848:** the Mexican–American War (see Book 5 of *A History of US*)

**1848:** Zachary Taylor elected 12th president

**1848:** Wisconsin becomes the 30th state

# More Books to Read

At the end of a history book, most writers make a list of books for further reading. I decided I would tell you about some books that I *love* to read:

**Joan W. Blos,** *A Gathering of Days,* Scribner's, 1979. Between 1830—when Catherine Hall, a New Hampshire farmer's daughter, is 13—and 1832, when she leaves home for the first time, she keeps this journal of the ordinary and extraordinary events of her hard yet happy life. You will feel you know Catherine by the end of this lovely book.

**Kevin Conley,** *Benjamin Banneker, Scientist and Mathematician,* Chelsea House, 1989. This book is much more interesting than most biographies of famous people and is stuffed with fascinating facts about slavery, inventors, and technology in colonial and Federal times.

**Alice Morse Earle, ed.,** *The Diary of Anna Green Winslow: A Boston School Girl of 1771,* Corner House, 1974. This is the kind of book I like to read. It is a real diary.

**Paul Fleischman,** *Path of the Pale Horse,* Harper Collins, 1983. Lep Nye is apprenticed to the local doctor in his little Pennsylvania town in 1793, when yellow fever grips Philadelphia. Lep and Dr. Peale go to the city to help in the awful epidemic. This is an exciting story, sometimes creepy, sometimes funny, by a good writer.

**Paul Fleischman,** *Townsend's Warbler,* Harper Collins, 1992. John Kirk Townsend and Thomas Nuttall were real naturalists who went west on the Oregon trail in 1834, examining and preserving plants, birds, and insects. This is a telling by an excellent writer (see *Path of the Pale Horse,* above) of Townsend's own journal of their travels. It has good pictures and maps, too.

**Russell Freedman,** *An Indian Winter,* Holiday House, 1992. Beautiful paintings by the artist Karl Bodmer illustrate this excellent book about the life of Mandan and Hidatsa peoples in the 1830s—the northern Plains Indians whom Lewis and Clark and George Catlin also met and described.

**James Cross Giblin,** *George Washington: A Picture Book Biography,* Scholastic, 1992. Michael Dooling's pictures make this book about the childhood and grownup career of the first president quite special and easy to read as well.

**Suzanne Hilton,** *A Capital, Capital City, 1790–1814,* Atheneum, 1992. This interesting book, illustrated with photographs, tells all about the beginnings of Washington, D.C.

**Katherine Paterson,** *Lyddie,* Lodestar, 1991. Lyddie Worthen is 13 when her mother gives up the struggle to keep up their poor Vermont farm without her husband. Lyddie heads for Lowell, Massachusetts, and the cotton mills. She earns her living and learns to read and write well; she suffers from the bad working conditions and witnesses the millgirls' efforts to get them improved. Lyddie is a splendid, real person and this is a wonderful book.

**Gary Paulsen,** *Nightjohn,* Delacorte, 1993. One slave secretly teaches another slave to read in this beautiful, sad book. It has great illustrations.

**Ann Rinaldi,** *Wolf by the Ears,* Scholastic, 1991. Seventeen-year-old slave Harriet Hemmings belongs to Thomas Jefferson. She lives a life—very unlike that of most slaves—at Monticello itself, and must decide whether to stay a slave in a golden cage or escape and find her freedom. This is a complicated but riveting story.

**Victoria Sherrow,** *Huskings, Quiltings, and Barn Raisings,* Walker, 1992. Events like these are called "work-play parties." Neighbors in early America would come together for a day or a weekend to accomplish big tasks—like building a barn or sewing all the pieces of a quilt—and have lots of fun at the same time. This book shows how the jobs got done and tells about them in people's own words.

**Michael O. Tunnell,** *The Joke's on George,* Tambourine, 1993. This is a short book, with nice pictures, about a real joke that the painter Charles Willson Peale once played in his museum on George Washington himself. It is easy to read.

# Index

# Picture Credits

engraving from *The History of the Indian Wars*, 1846, Chicago Historical Society, Charles F. Gunther Collection; **67**: Henry Inman, *Tenskwatawa, the Prophet*, after Charles Bird King, National Portrait Gallery, Smithsonian Institution; **68**: Eli Lilly, Indianapolis; **70**: Library of Congress; **71**: National Portrait Gallery, Smithsonian Insitution; **72**: Bureau of American Ethnology, Smithsonian Institution; **73 (top)**: New York Public Library; **73 (bottom)**: George Catlin, *Ball-play of the Choctaw*, National Museum of American Art, Smithsonian Institution; **74**: American Antiquarian Society; **76 (top)**: Library of Congress; **76 (bottom)**: Gilbert Stuart, *Dolley Madison*, ca. 1804, Pennsylvania Academy of Fine Arts, Harrison Earle Fund Purchase; **77**: Anne S. K. Brown Military Collection at Brown University; **78**: John Bower, *A View of the Bombardment of Fort McHenry,* aquatint, ca. 1815, Chicago Historical Society, gift of Charles B. Pike; **79**: National Portrait Gallery, Smithsonian Institution; **80**: New Orleans Museum of Art, gift of Colonel and Mrs. Edgar Garbish; **81 (top)**: Library of Congress; **81 (bottom)**: Department of Armed Forces, National Museum of American History, Smithsonian Insitution; **82**: West Point Museum; **83 (top)**: New York Public Library; **83 (bottom)**: Gibbes Museum of Art/Carolina Art Association; **85**: New York Public Library, Prints Division; **86 (top)**: Library of Congress; **86 (bottom)**: courtesy Mrs. Henry L. Mason, Boston (photo: Society for the Preservation of New England Antiquities); **87 (top)**: from William Tatham, *An Historical and Practical Essay on the Culture and Commerce of Tobacco*, 1800, Library of Congress; **87 (bottom)**: William Thornton, *Monticello*, Langdon Clay, New York; **88**: New York Public Library Picture Collection; **89**: New York State Historical Association; **90**: Gilbert Stuart, *John Adams,* National Museum of American Art, Smithsonian Insitution, Adams-Clement Collection, gift of Mary Louisa Adams Clement in memory of her mother, Louisa Catherine Adams Clement; **91**: New-York Historical Society; **92**: Tennessee State Library and Archives; **94 (top)**: National Portrait Gallery, Smithsonian Insitution; **94 (bottom)**: Library of Congress; **95**: James T. White & Company; **96**: Yale University Art Gallery, Mabel Brady Garvan Collection; **97 (top)**: Smithsonian Institution; **97 (bottom)**: *Ladies' Dressmaker*, Historical Society of Pennsylvania; **98 (top)**: Old Sturbridge Village and Lowell Historical Society; p. **100 (inset)**: David C. Hinman, *Eli Whitney,* ca. 1847, Chicago Historical Society; **101**: American Antiquarian Society; **102 (bottom)**: Bureau of Public Roads; **103**: National Library of Scotland, Edinburgh; **104**: Glenbow Foundation, Calgary, Alberta; **105 (bottom)**: Metropolitan Museum of Art, Harris Brisbane Dick Fund,1941; **107**: Stokes Collection, New York Public Library; **108 (top)**: Independence National Historical Park; **108 (bottom)**: Virginia Historical Society, Richmond; **109**: Library of Congress; **110 (top)**: Metropolitan Museum of Art, Rogers Fund, 1942; **110 (bottom)**: *The Orukter Amphibolis*, Historical Society of Pennsylvania; **111 (top)**: Ironbridge Gorge Museum Trust; **111 (middle)**: American Association of Railroads; **111 (bottom)**: Museum of Science and Industry, Chicago; **112 (top left)** Ironbridge Gorge Museum Trust; **112 (bottom right)**: Smithsonian Institution; **113**: Ironbridge Gorge Museum Trust; **114**: detail of engraving in Thomas McKenney and James Hall, *The Indian Tribes of North America*, 1836–44, National Museum of Natural History, Smithsonian Insitution; **115 (bottom)**: New York Public Library; **115 (inset)**: engraving in Thomas McKenney and James Hall, *The Indian Tribes of North America*, 1836–44, National Museum of Natural History, Smithsonian Insitution; **116**: National Collection of Fine Arts; **117 (left)**: Thomas Gilcrease Institute of American History and Art, Tulsa, Oklahoma; **117 (right)**: State Historical Society of Wisconsin; **118 (left)**: New-York Historical Society; **119**: John Wesley Jarvis (attr.), *Black Hawk and His Son, Whirling Thunder*, 1833, Thomas Gilcrease Institute of American History and Art, Tulsa, Oklahoma; **120**: William L. Clements Library, University of Michigan; **121 (top)**: Manuscript Division, University of Oklahoma Library; **122**: George Catlin, *Black Hawk and Five other Saukie Prisoners*, 1832, National Gallery of Art, Washington, D.C., Paul Mellon Collection; **123**: *Delaware River Water Front at Walnut Street*, ca.1835, Historical Society of Pennsylvania; **124**: Library of Congress; **126**: Bureau of American Ethnology; **127 (top)**: New York Public Library; **127 (bottom)**: John Clymer, *United States Marines Penetrating the Everglades*, 1944–45, National Archives; **128**: New Haven Colony Historical Society; **129**: Schedule of the 1790 Census, Philadelphia, October 24, 1791, Chicago Historical Society; **130**: Benjamin Henry Latrobe, *Market Folks*, 1819, National Museum of American History; **130-131**: New-York Historical Society; **131** (right inset): *Hillsborough Recorder*, February 21, 1849, North Carolina Department of Cultural Resources, Division of Archives and History; **131** (bottom left): *Group of negros, as imported to be sold as Slaves*, from J. G. Stedman, *Narrative of a Five-year's expedition against the revolted negros of Surinam...from the year 1772 to 1777* (1796), Library of Congress; **132**: Virginia Historical Society; **133 (top two)**: Library of Congress; **133 (bottom)**: note from Hannah Harris to Robert Carter, April 5, 1792, Chicago Historical Society; **134**: Benoît Louis Prevost, *Henry Laurens,* after Pierre Eugène du Simitière, Metropolitan Museum of Art, bequest of Charles Allen Munn, 1924; **136 (top)**: Lebanon Plantation slave cabin, Averasboro vicinity, Harnett County, Historic Sites Section, Division of Archives and History, North Carolina Department of Cultural Resources; **136 (bottom)**: Massachusetts Historical Society; **137**: Mason and Maas, *Captain Paul Cuffe*, after a drawing by John Pole, 1812, National Portrait Gallery, Smithsonian Institution; **138 (top)**: Museum of Art, Rhode Island School of Design, Providence; **138 (middle)**: Moorland-Springarn Research Center, Howard University, Washington, D.C.; **138 (bottom)**: Rembrandt Peale, *Absalom Jones*, Wilmington Society of Fine Arts, Delaware Art Museum; **140**: Library of Congress; **141 (top)**: The Old Print Shop, Inc., New York; **141 (bottom)**: South Carolina Historical Society; **142**: Taylor, *The American Slave Market*, 1852, Chicago Historical Society; **143**: Dr. Lloyd E. Hawes; **144**: New York Public Library; **145 (left)**: New-York Historical Society; **146 (top)**: Boston Athenaeum; **146 (bottom)**: Library of Congress; **147–149**: New York Public Library; **152**: Library of Congress; **153–155 (top)**: National Portrait Gallery, Smithsonian Institution; **155 (bottom)**: Library of Congress; **155 (right)**: Boston Athenaeum; **156 (top)**: National Portrait Gallery, Smithsonian Institution; **156 (bottom)**: Library of Congress; **157 (top)**: National Portrait Gallery, Smithsonian Institution; **157 (bottom)**: Library of Congress; **158**: silhouettes by Auguste Edouart, National Portrait Gallery, Smithsonian Institution; **159**: Dartmouth College; **160**: Chester Harding, *Henry Clay*, ca. 1820, Amherst College; **161**: National Portrait Gallery, Smithsonian Institution; **162**: Library of Congress; **163**: New York Public Library; **164**: Samuel Morse, *Robert Young Hayne*, ca. 1820, Governor's Mansion, Columbia, South Carolina; **164–165**: City of Boston Art Commission; **166, 167**: National Portrait Gallery, Smithsonian Institution. **175**: National Museum of Natural History, Smithsonian Institution.

*Sequoyah and his Cherokee alphabet*

# A Note from the Author

*An author is someone for whom writing is more difficult than it is for the rest of us.*

—THOMAS MANN

Do you know what those words mean? Well, I do, and I'm not going to tell you. I've done a lot of explaining in this book, but this is one thing you are going to have to figure out on your own. I will tell you that Thomas Mann, who was a 20th-century German writer (and a very good one), knew exactly what he was talking about.

Now that you are on your own with that quotation, I will tell you something—I will tell you about making books. It is *much* more complicated than I realized. Writing is only part of it. It takes many people to make a book. Here are some words about a few of those who have helped make these books what they are.

Tamara Glenny. She is my editor. Tamara is an Englishwoman, but, since her husband and children are American, I try to overlook her allegiance to the home of George III (it happens to be the home of William Shakespeare, too). Tamara checks my spelling and my grammar, but mostly she checks my thoughts and ideas. If I'm not clear about something, Tamara lets me know. She also puts everything—words, pictures, and design—together. That seems like magic to me. Good editors are rare. Tamara is extra good.

Mary Blair Dunton. Mary is the picture finder. Where does she find pictures? (In libraries, books, museums, and I-don't-know-where-else.) What do you think of the pictures in this book? I think they are great. Besides being a good picture researcher, Mary, as you can see, has a sense of humor. And besides all that, she comes from Virginia's Northern Neck, which is where George Washington was born, and Robert E. Lee, and Robert Carter III.

Mervyn Clay. Mervyn sits at a big computer and moves words and pictures about; sometimes he writes funny messages in the pictures. (Too bad you can't see them—you'd laugh.) He is the one who decides what size the pictures should be, and where they should go, and what the whole book should look like. Do you think this book looks like most books you read? I don't, and I am glad of it, and that is because of Mervyn. He is the book designer, and he is terrific. Unfortunately, Mervyn has a problem: he never liked history when he was in school—and he hated memorizing dates. This is more than an ordinary problem; it is a *phobia* (FO-bee-ya—from the Greek word for fear, *phobos*; it means an exaggerated or unreasonable fear). These books seem to be helping Mervyn with his problem. I think he is beginning to like history. But dates—he is still phobic about them. We are all trying to help him deal with his phobia.